MATH
for
HUMANS

Teaching Math
Through
8 Intelligences

By Mark Wahl

LivnLern Press
Langley, Washington 98260

Math for Humans: Teaching Math Through 8 Intelligences

ISBN 0-9656414-8-1

Pen and ink illustrations, layout and design, computer graphics and cover design by Dove and James Graphics, Clinton, Washington. Some computer graphics by P. Horbett Graphic Designs, Seattle, Washington. Grateful acknowledgement to Drew Kampion for editorial expertise.

Printed on 30% post-consumer recycled paper.

Library of Congress Catalog Card Number: 99-94575

The author can be reached by e-mail at **mathman@MarkWahl.com**.
Click to **www.MarkWahl.com** to find other math books, services and resources.

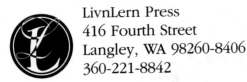

LivnLern Press
416 Fourth Street
Langley, WA 98260-8406
360-221-8842

"A mathematician, like a painter or a poet, is a maker of patterns. If [the mathematician's] patterns are more permanent than theirs, it is because they are made with *ideas*... The mathematician's patterns, like the painter's or poets, must be *beautiful*; the ideas, like the colors or the words, must fit together in a harmonious way..."

Mathematician G.H. Hardy

"...the desire to arrive at logically connected concepts is the emotional basis for [my] rather vague play with...more or less clear images which can be 'voluntarily' reproduced and combined. ...In my case [the images are] of visual and some of muscular type...analogous to certain logical connections one is searching for. ...This combinatory play seems to be the essential feature in productive thought — before there is any connection with logical construction in words ...which can be communicated to others."

Albert Einstein

"*The alternation of tension and relief* is a universal emotion. In reading any form of serious mathematics, we experience alternately perplexity and illumination. ...This, reaching the deepest levels of feeling, gently stimulates the aesthetic sensibilities. The effect is found in music when the alternation in (e.g.) a hymn tune of the dominant and the tonic — tension and relaxation — contributes to the beauty of the tune."

Mathematics and Physics Professor H. E. Huntley

To my daughter Hannah and the hundreds of other youths who have taught me how young people learn math. And to my wife Flay who has taught me to seek balance between my mathematical-logical and my intrapersonal intelligences.

TABLE OF CONTENTS

Section B. *14 Activities (over 90 sub-activities) for the Student*

Section C. *Appendix*

Preface

THE RATIONALE FOR THIS BOOK

In 1967, as a graduate student in math education at the University of Maryland, I was teaching a challenging course in the "New Math" to prospective elementary teachers. At one point I found myself with a weeping coed in my office. I felt extremely awkward as we tried to resolve her struggles with mathematics. My mentor and supervisor, an accomplished math educator named Helen Garstens, warned me (as the surrogate mother she loved to be) not to get too pulled in by the student because tears are just tears and sometimes they're used as a form of grade leverage. My main recollection from that event is that math and tears seemed a baffling mix from my perspective — having just earned my masters degree in the bloodless higher realms of abstract math.

This "tears and math" conundrum was filed away until about 1976 when Sheila Tobias popularized the phrase "math anxiety" in Ms. Magazine thus giving millions a name for feelings they had been harboring and trying to forget. I recall that at that time there was a tremendous national outpouring of recognition. The math education establishment took notice, has periodically done some research over the last two decades, and occasionally still makes references to this phenomenon, but not much has been done to directly attend to the math-tears connection.

For me, Tobias' overt recognition of the phenomenon of math anxiety marked the crumbling of a huge wall in myself separating mathematics from the personal realm, a wall that most societies of the world still train students to maintain. Ever since, I have become increasingly aware of the influence of the personal and "feeling" realms on the learning of mathematics. I've seen at close range how art lessons single-handedly improved children's math performance, and how a young girl's recognition of her suppressed inner feelings directly raised her math grades.

Because of this link to the personal, I was led to explore the historical, psychological, philosophical, and artistic dimensions of mathematics. I taught a course at the University of Denver called "Mathematics in Art and Design"; I also conducted numerous workshops on math anxiety, and eventually published my well-accepted book, **A Mathematical Mystery Tour,** with real-life activities for young people that explore the mysterious links of number with art, nature, history, and even spirit.

In the late '80s I encountered Howard Gardner's theories of the Multiple Intelligences. This provided a framework for my understanding of the various humanistic links I had made with and to mathematics. My appreciation of Gardner's theory has continued to blossom since then, and has now resulted in this book, which merges his theories, my experience, some other theories of education, and the current trends in the math teaching establishment.

Since 1967 I have developed curriculum and taught mathematics at levels ranging from primary to college and adult, in standard and very experimental class settings, including home-schooling contexts. Another major laboratory for me has been the extensive observation and instruction I've given to many students of all ages in one-on-one sessions where I am able to obtain a 90+% success rate for rapid turnarounds in attitude and performance (often requiring as little as three or four hours of mentoring). In addition, I have also conducted teacher inservice training for 17 years on many of the ideas I take up in **Math for Humans**.

Meanwhile, over the course of all those years I've watched pedagogical ideas in math teaching come and go, from the "New Math" of the '60s, to the math labs of the '70s, to the thinking skills units of the '80s. Now math educators are in a constructivist era; they're exploring cooperative learning (terms I will explain in my early chapters). Though few math education theorists are currently referring to the Multiple Intelligences, this will probably become part of the jargon of the 2000s and beyond.

Another growing influence on teaching is that of *inclusive education*. The 1997 revision of the Individuals with Disabilities Education Act (IDEA) increases the rights of children with various disabilities to be educated in a general classroom setting. This involves creating Individualized Education Plans (IEPs) that require a team including the general classroom teacher to specify adaptations for the child with disabilities. In math this includes seeking approaches that tap the strongest intelligences of that child. **Math for Humans** is a perfect resource for the inclusive classroom teacher. (Inclusive-thinking teachers usually arrive at the fact that *all students* need adaptations!)

Time and again I have encountered teachers, curricula and schools that are very innovative in other subjects, but merely "settling" for mediocre materials and techniques to communicate daily math lessons to students. I've seen teachers of the lower and middle grades with humanities backgrounds regularly generating exciting lessons in language and social studies while conducting math periods filled with simple blackboard "telling" and worksheets.

As a result of all the influences and observations I've described, you now hold in your hands the seeds of a new way of teaching and learning math. Believe me, there are already *plenty* of books

The theory and Activities in Math for Humans bring the teacher or homeschooling parent to a higher level of effectiveness in multiple ways.

that offer math activities and pedagogy, but **Math for Humans** will gently guide you to a broader and more flexible approach than you will find in most other resources. It certainly won't exclude the use of those resources. In fact, its expansive principles will give you criteria for sifting the best ideas from them. If you pursue, in your own way, with your own best gifts, the philosophy, strategies and concepts presented in this book, you will arrive at a new depth of communication and experience that you, your students, and their parents will cherish.

The touchstone for my approach, the theory of Multiple Intelligences (MI) of Howard Gardner, is gaining more adherents each year. This theory speaks of each person as possessing not just *one* but at least *eight* different intelligences to learn with. You may have heard of MI Theory and read up on it. You may have taken workshops using it, or tried it in some of your lessons already. Nevertheless, resources relating MI to the broad spectrum of *math* teaching have been rare or non-existent — yet more motivation for me to write **Math for Humans**.

Whether or not you've encountered MI before, in these pages you'll find ample opportunity to begin schooling yourself in MI as the theory specifically relates to *math* learning and teaching. I predict that before long you'll be creating your own activities, which will reflect your increasing understanding of this new perspective.

As I mentioned earlier, **Math for Humans** doesn't stop with MI. I've done my best in these pages to introduce you to some other key theories and strategies, too. I've also shared the benefits of my own experience, in hopes this will further augment your math teaching. All of these components fit nicely under the broad MI umbrella.

For many of you, what will evolve from using this multifaceted book is a change of mindset; you will begin to see with new eyes. This richer approach to teaching and learning math will facilitate a commitment to fun, quality, insight, versatility, and feeling. It will engender an expanded understanding of how children learn, leading you to celebrate the genius in each and every child. Metaphorically, this is a change from black and white to color in math teaching.

The recent Third International Mathematics and Science Study found that U.S. math teachers are among the busiest in the world, and the National Commission on Teaching and America's Future found that most elementary school teachers have minuscule preparation time (8.3 min.) per class. Given your busy schedule you probably don't have hours to dedicate to achieving total transformation by next week! The good news is that **Math for Humans** provides many methods of *easing* you into the MI approach with small investments of time. As you continue, you will find that, day by day, MI will whet your appetite for more. And what's even better, as you grow more comfortable with MI techniques they will save you instructional time in many ways.

You have some choices of how to begin. Be sure to read the introduction. It will give you an overview from which to choose a pathway into the MI math approach. From its chapter descriptions you can choose how to read up on the conceptual elements that underlie the MI instructional paradigm shift (yes, that's what we're talking about here).

You might choose a more *experiential* way, letting the MI model rub off on you as you explore with your students actual Activities from the black-line masters in the second half. Using the extensive Teacher Notes I've provided for each Activity, you'll learn nuances of MI theory as they relate to specific math experiences. And, last but not least, you can select from a large variety of MI strategies to build alternatives into your regular curriculum every day. Soon you'll probably feel the urge to plan some Multiple Intelligences lessons yourself!

I encourage your feedback on these theories and Activities. Tell me what works and what needs to be improved. Send your comments and questions to me at **Mark Wahl Learning Services,** 416 4th Street, Langley, WA 98260, or e-mail them to me: mathman@markwahl.com.

Happy Teaching!

Mark Wahl

Introduction

THE WHY AND HOW OF THIS BOOK

As a math learning specialist I have had considerable experience dealing with the pain of mathematics education carried by children, adults (former children), and many teachers (when they talk candidly). For years I've taught an "Overcoming Mathephobia" seminar for adults and teens. Boy, do I get an earful about scars from inappropriate tactics applied by well-meaning teachers! Math pain is an epidemic in our culture. Despite reform efforts by educators for the 40 years since the *Sputnik* satellite told the U.S. we were technically behind the Russians, it's still an epidemic. Of course, there are those kids who make it through our educational system just fine, and there are those insightful teachers who pull it off well, but it still amazes me how widespread math aversion remains. The question is, what can really start to turn this situation around?

I remember Tory, a second grader and a real "dolphin" (you'll meet these in the Learning Styles chapter!) — dreamy, artistic and sensitive. She was one of the most spatially-intelligent kids I have met — a budding artist. Her addition tables absolutely eluded her despite her teacher's best efforts (in a private school with low student-teacher ratio). Because of this, the flower that she was started closing her petals in self-protection. While I personally don't believe that forcing math tables onto young students whose natural math development curve is a bit delayed is at all necessary, I knew she needed to meet her parents', teacher's, and fellow students' expectations or she would whither.

Current progressive teaching wisdom would have dictated that Tory spend time with balance scales, cubes, number patterns, or games to learn the tables, but I decided to capitalize on her spatial-artistic intelligence instead. I asked her to make a picture incorporating the symbols of $8 + 7 = 15$ on a large index card. I requested she do this for about four other such math facts as well.

Tory returned happily presenting five cards covered with color — no white visible. Each was a picture in which could be discerned (sometimes with her guidance) the symbols of a math equation creating the outlines of trees, persons, beach towels, etc. I looked at one and asked her, "How much is $8 + 7$?" She couldn't answer. Oh, well, I thought, so much for a good idea. Then I said "It's the beach scene," and she immediately said "15." With such cueing I was able to have her more or less recall the other answers.

She begged me to assign her some more to draw, so I obliged.

She ended up with a deck of artistic flash cards that soon created answer associations for her without my having to provide scene cues. My logical-mathematical intelligence couldn't even figure out *how* her mind did this, but there was no need to — she was successful with her facts, and her math performance took off, thanks to her spatial intelligence.

I recall Kate who kinesthetically learned multiples by jumping rope when other means had failed; and Nate, who walked number multiplication answers in the sand, and Sheryl, who quickly learned when she listened to tapes of math facts made by her dad (an accomplished musician). And there are those many students from second graders to adults who rapidly embrace fractions as I relate my dramatic story of two bizarre animals that act out the fundamental meanings of numerator and denominator.

These are but a few examples of the effectiveness gained when a teacher maintains awareness of the Multiple Intelligences (MI). **Math for Humans** will give you the practical tools for using this theory in five ways:

1) Chapter 1, "The Theory of Multiple Intelligences: A Sketch from a Math Teaching Perspective," explains the key concepts of MI.

2) Other introductory chapters briefly describe theories and strategies that overlap parts of MI theory.

3) Chapter 3, "Seasoning Your Math Teaching With MI," provides ways of incorporating bits or large doses of MI into your regular fare.

4) The ready-to-use Activity masters and accompanying Teacher Notes are designed to instruct students (and you) through the Multiple Intelligences in action.

5) The "Resource Bibliography" will give you other books for learning about, and teaching with, a kaleidescope of materials and techniques related to all the intelligences.

About the two halves of *Math for Humans*

In the **first half** there are initial **theory chapters** that include:

➤ a sketch of the theory of Multiple Intelligences (MI),

➤ an overview of the *Standards* of the National Council of Teachers of Mathematics,

➤ a short tour of the hemispheres of the brain,

➤ a handy system of learning-style information,

➤ an introduction to cooperative learning, and

➤ some notes on assessment, which will enhance your math lesson planning and execution.

"*I'm Wily Worm and I'll be sure to put my two cents worth in these margins and the student Activities.*"

There are also special chapters on three chronic math challenges to teachers of several grades:

7. Dealing with Math Anxiety,
9. + - x ÷ Supercharged! — Power-packing the Under — standing of the Four Operations with MI, and
10. Sharpening the Math Facts.

There is a loaded and spicy "MI Seasoning" chapter that can be a gateway for your entrance into the world of teaching math through the Multiple Intelligences. It contains numerous ways to sprinkle (or pour) the Multiple Intelligences approach into all your daily math lessons. It's a small manual in itself that you can slowly absorb as you go along in your regular math teaching.

The **second half** of **Math for Humans** contains a total rewrite of some of the math activities I published over 20 years ago. In their earlier forms these Activities were used by thousands of teachers and homeschooling parents, and by hundreds of thousands of American and Canadian kids. I have used them with my own students over the years and have continued to hone and improve them as well as develop new ones. Some are well-known enrichment math ideas adapted with my own presentational spin. Several Activities are based on my original mathematical insights or on my original use of known data for instructional purposes.

The **Math for Humans** Activities have much to offer to a range of learners. Average and even struggling students of ages 8+ to 12+ (or beyond) will profit from them because of their multifaceted learning approaches and their adaptability to varying levels. They are excellent for mathematically-strong or gifted students because of the suggested extensions (see the light bulbs) and because they range from moderately easy to very challenging. The amount of teacher input can be varied according to the strengths and weaknesses of students.

The feedback from the older, simpler versions of these Activities has been very positive, so I anticipate that teachers and students will be even happier with my reformulated, expanded, and updated versions.

Here are some of the **features** of the Activities section:

> They are "Ready to go" Activities that teach math through the Multiple Intelligences, utilize the whole brain, and further the objectives of the Curriculum *Standards* of the National Council of Teachers of Mathematics.

> The Activities teach new math topics not ordinarily included in the curriculum (e.g., ellipses) and new approaches to old topics (e.g., working with large numbers in "Africa Counts").

"As Math Mole, I'm around to set a mood, to suggest ideas, and to entertain you and your students."

➤ The Activities richly *model* the way Multiple Intelligences can interact with math in the classroom and the real world.

➤ The Activities offer Teacher Notes that provide thoughtful theoretical and practical support for gleaning MI lessons from the activity. (e.g., "Translate this abstract activity into a musical mode by asking students to change number patterns to note patterns on the piano... ")

➤ There is a Finder Key before the Activities that allows teachers to choose Activities at desired levels of challenge, emphasizing certain intelligences or certain math skills. This can also help with individualizing tasks for exceptional learners.

➤ The Activity pages are graphically alive and carefully sequenced in order to hold student (and teacher) interest. Featured in the illustrations is "Math Mole," a cartoony model of persistence and sleuthing ability, who has many guises in the Activities, and "Wily Worm," his quarry, who contributes emphasis and clarifying comments.

➤ Rather than trying, as most math enrichment collections do, to hook the logical-mathematical intelligence only, the Activities seek to engage the other intelligences by using a wide variety of modalities.

➤ Real information about our world is inserted wherever possible so that student calculations yield meaningful, educational results instead of irrelevant busywork.

NOTE: In these Activities numerical facts are almost always presented in our Common **U.S. units**, rather than in the metric units now used widely in textbook problems. I do this very purposefully for U.S. students because the real-life number facts and the answers computed from them are quite striking when students view them from their everyday experience. (For example, U.S. kids never speak of a car's speed in km/hr despite the fact that their math books do. And which would you or they "feel" and re-member better, that lead weighs 11.3 grams per cubic centimeter or that a cube of lead two inches on a side weighs $3\frac{1}{2}$ pounds?) Feel free to "white out" these units and substitute metric.

Using U.S. Common Units in problems creates a link between math class and life, for most students using this book, something I want the Activities to exude. However, for the math teaching purists I hereby enter into the record that I believe that the metric system *can* and *should be* instructed at many other times in realistic ways so that the students can speak *both* measurement languages in this "bi-metrical" country and world.

Math for Humans should be thought of as a versitile tool for quickly improving daily math lessons <u>and</u> for long range teaching improvement.

Section A:

For the Teacher

Multiple Intelligences

THE THEORY FROM A MATH-TEACHING PERSPECTIVE

For the past 95 years, the classic IQ test of psychologist Alfred Binet, developed in Paris to predict school failure and success, has been the standard for the objective measurement of a person's level of "intelligence." One's IQ is considered by the average person to be the essentially unchangeable measure of the quality of his or her genetically-determined "hardware."

Personally, I'm quite happy I never found out what I scored on the test when I took it in school in the '50s. If I had known it was high, it might have inflated my estimation of my brain power and I might have attempted to "coast on my laurels," even possibly joining the ranks of what is called the "underachieving gifted." I might also have failed to broaden my learning in several areas because I would have taken my high score as a sign that I was destined to be an ivory-tower academic or technologist.

If, on the other hand, I had discovered my IQ score was low, or even just an average 100, I might have branded myself "average" or "slow." I might have immediately acquired an alabi for not striving for excellence, excusing myself from various academic challenges, like the geometry I took in 10th grade with a class of seniors, or the General Class ham license I got at age 15. Or worse, I might have carried convictions of my ordinariness and my mental deficiencies into adulthood.

In fact, a low score on Binet's test (or a high score, for that matter) can be the result of any number of variables that undermine its purported "objectivity." Some influencing factors can be physical and emotional things like diet, sleep, nutrition, and home life. But, more important for this discussion, the test-taker may have a different primary mode of intelligence that is not even being tested.

The point is that the IQ number "doth not the person make." IQ tests measure only a narrow slice of **two** (verbal and logical-mathematical) of the (at least) **eight** intelligences identified by Howard Gardner (*see Resource Bibliography: Gardner*). The work of Isreal's Reuven Feuerstein, yielding huge competency gains in the mentally-handicapped, and many other findings have demonstrated that Binet's IQ number is not some kind of fixed "hardware" measure. Even when fairly low it can be altered by training, and it can change and grow as stimuli increase in one's environment (increasing the ganglia interconnecting brain cells).

Think for a moment. If a friend says to you that she met a *very intelligent* person, what would you picture? Probably someone verbally or logically adept. You are reflecting *our culture's* norm that verbal facility and/or logical-mathematical facility are the defining factors of what we call intelligence.

In the spiritual and artistic culture of Bali or in the ocean-navigating cultures of Micronesia, a high-IQ "nerd" would be considered fairly dull-witted. In our own culture, a movement virtuoso like Fred Astaire or a musical powerhouse like Ray Charles may not have rated particularly high on IQ tests, and so might not be thought of as *intelligent,* just "talented." But a verbally *talented* Tolkein and a logically-mathematically *skilled* Bill Gates would clearly be considered "intelligent" in our Western culture.

The Essence of Gardner

Howard Gardner's extensive research, first outlined in his book *Frames of Mind* (1983) showed that there are at least seven semi-independent ways of "solving problems and fashioning products" that are equal candidates for being called "intelligences." In 1997 he documented on eighth — the Naturalist Intelligence.

Gardner identified and examined unique timelines for the development of these intelligences in children. Further, he noted the independent breakdown of each in conditions of localized brain damage. He marvelled at highly developed but single-intelligence manifestations in autistic savants and prodigies, and he documented different intelligence emphases in other cultures. Additionally, he observed and recorded the non-transference of these intelligences — development in one intelligence does not automatically produce better performance in another.

The semi-independent modes of "solving problems and fashioning products" (i.e., intelligences) that Gardener identified are these:

Musical Intelligence — formerly considered just a talent, this is a full-fledged mental modality on a par with, say, the verbal intelligence. Somewhat localized in the right brain, the musical intelligence's grasp of subtlety and complexity is virtually unlimited. It includes the ability to make music or rhythm and/or to feel or hear it in ways that evoke pleasure, surprise, moods, and a type of mental satisfaction independent of rational thinking.

A musically-intelligent student may not necessarily play an instrument. Perhaps, he just learns commercial jingles rapidly or loves to listen to music. He might benefit from rhythmic recitation or songs with math information . He might be made alert or calm by a piece of music before working. Such a

student will benefit considerably from hearing the music of a culture that math information comes from.

Logical-Mathematical Intelligence — Used as a basis for IQ tests, this intelligence underlies the deductive methods of science and law. Its way of working is not just through symbols and syllogisms, however, but is often initially non-verbal — a *sense* of how causes are related through a series of steps that might later are put down in symbols or soundly reasoned words. Thus, logical-mathematical intelligence does not merely use mechanical methods. It is an intuitive *art* too. It draws from *several* areas of the brain.

Students with high logical-mathematical intelligence are often fascinated with patterns in numbers or with science and will pursue ideas far beyond their apparent utility. Not just an isolated facility, this intelligence finds application in many fields: biology, music, sports, art, ecology, politics, and law.

Bodily-Kinesthetic Intelligence — Strongly exhibited by dancers, athletes, and craftspeople, bodily-kinesthetic intelligence is used by many to gain insight, solve problems, and process information in other areas. For instance, a violinist uses it to control the bow even though melodic decisions are made by his or her musical intelligence. Bodily-kinesthetic intelligence is governed by areas near the center of each hemisphere of the brain and by the cerebellum deep in the back of the brain.

Students with more of a bodily-kinesthetic intelligence may learn by being "walked through" math procedures, having a "hands-on experience," getting a "gut feeling," and "going through the motions." They may be restless just sitting and listening; they may half mangle math papers. Bodily-kinesthetic students may learn more from body language and gestures of the teacher than from her words. Einstein felt for a body sense of how two physical variables related *well before* he attempted to put anything into mathematical equations.

Interpersonal Intelligence — Residing strongly in the frontal lobes of the brain, this is the capacity to perceive the seemingly hidden temperaments, motivations, needs, agendas, and intentions of others. Most of our lives are filled with interpersonal relationships; some of us are more agile at problem-solving in this arena than others.

Children with strong interpersonal intelligence may have excellent group skills, show empathy for others and/or

Gardner's conclusion: If we insist that some of these are just skills or talents, then we must concede that all are. If some are "intelligences," then all are.

demonstrate leadership, and be continually drawn into social situations. They may acquire friends easily and even obligingly alter their personae with each. In math they will usually respond to cooperative learning and cross-cultural lessons; they may be stimulated in math by learning the history and anecdotes behind the material they are studying.

Spatial Intelligence — Localized in the right brain, this intelligence often works in concert with vision, but blind persons exhibit it, too. Spatial intelligence is a fantastic aid to problem-solving in many areas: navigation, costuming, sculpture, engineering, room decoration, and computer (cyber-*space*) information management. A minority of people are "spatially challenged" — they get disoriented trying to unthread a map, turn the letter F upside down and backwards in space, or put together a machine they just took apart. But even they often benefit from simple use of spatial cues in the information they're studying (e.g. outlines, layouts, arrows).

Addressing students' spatial intelligence is perhaps one of the strongest aids in mathematics education. Diagrams, pictures, manipulatives, and games make mathematics clear to many.

Intrapersonal Intelligence — Also localized in the frontal lobes of the brain, this intelligence shows itself in the ability to create and use an accurate, undistorted model of oneself — knowing one's own "feeling life," thought dynamics, values, and personal traits. Someone strong in intrapersonal intelligence might be considered "thoughtful" or "sensitive," perhaps introspective. She might have much more than a perfunctory answer to the question "How are you today?" but may or may not have a great deal of *interpersonal* or *linguistic* skill for conveying the answer. A student strong in this intelligence would happily examine his own thought process in successfully arriving at an answer and would also meaningfully discuss his fears about math.

Linguistic Intelligence — Residing in Broca's and Wernicke's areas of the brain, linguistic intelligence is exemplified in poets and other authors, but it is a strength of many people who are comfortable in the universe of words, reading, or writing. These individuals are often articulate, having a sense of humor. Deaf children who invent their own signs are likely showing this intelligence at work. A linguistically-strong child with average math skills may do fairly well on an IQ test.

Math teaching can draw heavily on linguistic intelligence by encouraging children to put into words what they see as patterns and procedures. A careful choice of a few accurate

key words (instead of a stream of technical teacherese) to explain an idea can capitalize on the verbal capacity for clarifying concepts, and it can prevent the predictable misunderstandings that later need correction. Also, the origins and definitions of math terms can enrich a presentation for students who possess a strong linguistic intelligence.

Naturalist Intelligence — This intelligence has to do with observing, understanding and organizing patterns in the natural environment (plants, animals, rocks, and natural features) an ancient key to survival. Tom Brown, known as "The Tracker" can distinguish minute indicators in any footprint that indicate even the emotional state of the walker, whether animal or human! Even identifying brands and models of cars, analyzing fingerprint variations, or spotting pathologies in x-rays uses the naturalist parts of the brain. In the math "environment," attributes of fractals, unique features of graphs, and visual patterns created by numbers and geometrical designs all tap this brain capacity. The Fibonacci Number patterns even give us deep insight into natural forms. (For instructing this see the Resource Bibliography: Wahl.)

A major premise shared by Gardner, and this book, is that *a teacher should more often attempt to tap processes that are strong and natural in the learner than consistently demand production from a weaker intelligence mode.* Without understanding this fundamental dictum, schools and educators find themselves in classic no-win situations, as evidenced by the child who's constantly tipping out of his chair and throwing spitwads, using his strong kinesthetic intelligence for trivial ends, while the teacher is trying to verbally bang things into his weaker logical-mathematical intelligence, to no avail.

The times tables, for example, can be taught more effectively through several intelligences rather than relying on the traditional flash cards, printed square table, and timed tests that appeal to some linguistically- and spatially-intelligent students. For a variety of Multiple-Intelligence options that rapidly teach the math facts, see Chapter 10, "Sharpening the Math Facts."

Building on this fundamental structure, the Teacher Chapters and Activities in **Math for Humans** direct the teaching of mathematics toward consciously utilizing each of the intelligences as much as possible. This also means truly including the logical-mathematical intelligence that too often gets underplayed during instruction in favor of secretarial copying and rote imitation.

I hope you have been stimulated by reading this summary to look further into the eye-opening theory of Multiple Intelligences. Consult the Resource Bibliography in the Appendix to locate fascinating references for its further study.

Direct Teaching vs. Constructivism

NCTM STANDARDS AND THE MATH WARS

In 1989, the National Council of Teachers of Mathematics (NCTM) set out to enumerate a universal set of math teaching standards. They refined and honed them thoroughly in the 90s then updated them in 2000, calling them the *Principles and Standards for School Mathematics*, (*P & S*). (See NCTM.org for current versions of them.)

The universality of the *P & S* has been contested at times by those that feel they *underemphasize* math's mechanical procedures and the practice required to perfect them. In California, Texas and other states, where too many students still score low in the No-Child-Left-Behind testing, college math professors, parents and some teachers have started heated debates (coined "Math Wars") about texts espousing *P & S* methodologies. The critics strongly favor the more traditional and basic "direct teaching" with drill and practice of computational skills while critiquing the *P & -influenced* texts as "math lite." They accuse the *P & S* of espousing "soft-headed, feel-good methods that dumb down the curriculum." They assert that skills "discovered" by students are often not clarified by teachers or books and they are neither formulated for retention nor practiced to mastery. They add that follow-through from "experiences" is inconsistent.

Key: Mix constructivism and basic practice

On the other side, boosters of the *Principles and Standards* include many math educators, and the NCTM, who see "direct instruction" often as "giving the right formulas to produce the right answers," with teacher as "teller" and student as "listener" and "practicer." They warn that true analytic thinking, understanding, problem solving, and application ability go out the window.

To them a teacher is a mediator who guides experiences, investigations, discussions, and solutions of provocative problems, then facilitates the construction of ideas and methods from them. This overall approach is called "constructivism." Proponents cite impressive improvements in math scores and motivation where trained teachers use these methods.

A goad for reform is the documented widespread decline in math scores of U.S. students using either kind of approach.

Commonly scores progress from a relatively strong performance in fourth grade to moderate in eighth to abysmal in high school. Both sides blame the others' pedagogy and decry the lack of adequate teacher training.

Bill Riley, US Education Secretary, said "we are suffering here from an 'either-or' mentality." I concur: there is room for praise and criticism in both approaches. I have seen constructivism implemented inefficiently and dogmatically, resulting in watered-down skills and confusion about learning goals; I have also encountered boredom, shallow understanding and robotic responses to math situations where traditional texts and pedagogy prevail. Neither is a silver bullet; neither a complete dud. A skilled teacher can make both work; she must adjust teaching strategies to a range of student needs, teacher temperament, and class "personality." The MI approach developed in this book serves both strategies: enhanced concept development, drill, motivation, and real life applications.

The basic direct instruction model is familiar to most, so here is more on the constructivist *P & S* approach. Constructivism includes: working from big concepts to details, rather than the reverse; using primary sources of data rather than simulated information for problems; treating students as thinkers with emerging theories rather than as "blank slates" to be written on; interweaving assessment with the teaching and student activities rather than after them.

Let's use fraction multiplication to illustrate the difference in approach. In traditional classrooms, it is considered "a piece of cake" to teach a student to "multiply numerators and denominators and cross-cancel first if possible." Voilá! Right answers! Unfortunately, asking the student "Make up a real-life story problem needing the result of 2/3 x 9/10 and explain what the answer 3/5 tells you," earns a blank stare. Asking what the 2 on top and the 3 on the bottom do even earns confusion. Asking even what multiplication is in this context gets a vague response. (See Chapter 9 for an in-depth look at operations.) Basically, we have a blind robot getting "right" answers.

In contrast, the *P & S* might lead off with a hands-on exploration of fractions and pieces of fractions. The teacher might ask for answers from a calculator when 8 is multiplied by ever-smaller numbers, descending to fractions below 1. Patterns or trends would emerge from discussion: smaller-than-1 multipliers give answers smaller than 8. Also, fractions might be modeled on a group of 30 chips, showing that 9/10 of them (27) would be first found, then 2/3 of that (18) would be found next, and 18 is 3/5 of 30. Additionally, a rectangle with 10ths ruled vertically and thirds ruled horizontally illustrates with shaded areas the 3/5 answer and why multiplication of numerators and denominators is done.

An overview of the NCTM Standards

Just to give you a sense of the breadth of the *Principles and Standards*, the rectangle contains a collage of several skills, concepts, and processes advocated by them. You will naturally learn them in a hands-on way as you try the Activities in the second half of **Math for Humans**, learning from the rich assembly of teacher instruction and support information following each Activity group.

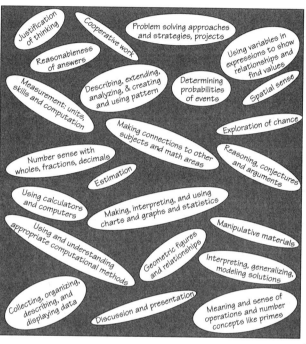

Another snapshot of the scope of the *Principles and Standards* is given by a list of their ten main topic/skill areas to be developed for all levels from K-12. Of course, each of these areas must be cultivated at levels appropriate to student age and ability, growing in depth and breadth through the grades. I'll put seed comments by each to help clarify them:

1. Number and operations [*Concepts, estimation, mental math, computation*]
2. Patterns, functions and algebra [*Number behaviors, patterns, interrelationships*]
3. Geometry and spatial sense [*Shape attributes, interactions, construction, space*]
4. Measurement [*Standard units, conversion, estimation, area*]
5. Data analysis, statistics, probability [*Gather, distill data, predict outcomes*]
6. Problem solving [*Investigate, strategize, solve*]
7. Reasoning and proof [*Infer, hypothesize, justify result*]
8. Communication [*Retrieving information, clearly reporting results*]
9. Connections [*Between math areas; to other subjects, real life*]
10. Representation [*Math notation, graphs, charts, diagrams*]

Algebra is begun in the lower grades where it consists mainly of number pattern work and finding missing numbers in expressions.

Each of the ten areas above must be developmentally appropriate for the students' ages and should be approached in more sophisticated depth and breadth through the grades.

More on the Standards, Basics, and MI

The *Principles and Standards* have been criticized in the "Math Wars" for underemphasizing "drill and practice." They may have been trying to bend the twig toward constructivism since basic skill proficiency alone has too often been equated with math proficiency. To respond to demands for more "basics," the NCTM has recently created a set of "Curriculum Focal Points" for each grade that emphasizes its "Basic" math skills (see NCTM.org). Regardless of your curricular emphasis, traditional or constructivist, note once again that the Multiple Intelligences strategies in *Math for Humans* are extremely adaptable to teaching the facts and mechanical skills *as well as* fostering the constructivist processes.

Now about **fun**. Others have felt that the *Principles and Standards* are high on serious Logical-Mathematical work while being lower on fun and the enjoyment of mathematics. The *P & S* certainly leave it up to the teacher to "lighten up" the lessons and to pursue fun topics. Once again, the Multiple-Intelligences approach can help add to the perception of fun in the classroom. Try the Activities included in the second half of this book: to see how "fun" is incorporated and to see how to utilize the many "channels" through which students receive, process, and store information. The *Principles and Standards* include nothing explicit about teaching to and through the Multiple Intelligences, nor do they systematically attempt to include these multifaceted approaches. Broadening to the Multiple-Intelligences perspective will give you new vantage points, energy, color, feeling, and adaptability to add to the *NCTM's* fairly strong encouragement of the Logical- Mathematical Intelligence.

Finally, a couple of useful terms. It is well known that there are two main kinds of reasoning: **deductive** and **inductive**. Deductive reasoning is strongly emphasized in the traditional math curriculum. Its starts with a principle or rule then applies this logically to examples—like a typical practice homework assignment after a procedure has been outlined in class. In life, most regular jobs get rewards for following the rules and procedures.

Inductive reasoning is the reverse of this: it starts with a collection of examples; then a rule that governs the examples is developed. This kind of reasoning is initially felt to be harder to do. It is the kind that often gets more monetary rewards in life. It's the spotting of the pattern, the "connecting of the dots" those TV heroes, investigative reporters and CEOs get good bucks for. It's also the kind of reasoning demanded more by constructivist teaching and ultimately enjoyed more by many students (though sometimes not the "beavers"—see Chapter 5 on Learning Styles and Math). Many teachers would do well to try to incorporate more inductive reasoning into their lessons.

Children need to be challenged yet invited regularly to have fun in math.

Seasoning Math with MI

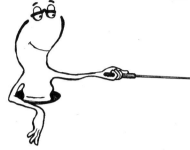

A LITTLE SPICE GOES A LONG WAY

Do you fit the following description? You've read about MI theory and heard about it from colleagues. You want to use it in your classroom to teach your regular curriculum. You have the activities in this book to try, but meanwhile you need to teach the *required* material. You feel daunted by how to start. You're intimidated by the all the changes and preparation it could entail. You feel pooped just thinking about it!

I feel bound by truth-in-marketing laws to honestly report that teachers who dive into the MI wave *do* start spending more energy on teaching, and they re-emerge on shores that look really different from the ones they left. I hasten to add that they feel energized by this paradigm shift.

But, relax. You can stick your toe in easily just to test the waters. Just enhance the flavor of your usual daily math instruction (or any other subject for that matter) with Multiple Intelligences seasoning and you'll notice your students beginning to perk up. That in turn will energize you, your class preparations, and your teaching day. You'll get used to that extra spicy edge. Careful! It'll suck you in! As you progress you may even feel a mysterious urge to dive in to major classroom and curriculum restructuring! Your use of the Activities in this book and your reading of selections from the Resource Bibliography in the back will give you further momentum.

So here are some ways to season any regular lesson with each of eight intelligences. You don't have to be an extremist by trying to use all for each lesson, or by trying to master all of the intelligences at once. Just pick, from the lists below, a few of your favorite techniques in any of the intelligences — or one technique from each intelligence — and try to work them into your day. Use them regularly, and they will become part of your approach. Add to them as you go along, and one day you'll look in the mirror and say, "I'm an MI math teacher!"

Seasoning with the Linguistic Intelligence

This intelligence has a natural carryover to math instruction. In fact, it's the one that has drawn a lot of focus in the Standards of the National Council of Teachers of Mathematics (See Chapter 2).

The 1998 *Standards* still express well the math-linguistic connection: "...The study of mathematics should include numerous opportunities for communication so that students can —

- reflect on and clarify their thinking about mathematical ideas and situations;
- realize that representing, discussing, reading, writing, and listening to mathematics are a vital part of learning and using mathematics...;
- model situations using oral, [and] written ... methods;
- use the skills of reading, listening, and viewing to interpret and evaluate mathematical ideas ...".

The *Principles and Standards* suggest several linguistic classroom strategies, and there are many more besides. *(For some of these, see the Resource Bibliography: NCTM, Burns, Myers.)* Here are some of my own, illustrated by examples, with suggested teacher language:

➤ After the student has experienced a pattern in his/her results, ask for a **generalization of the pattern.**

Teacher: "Try to communicate the pattern you just found by writing, in at least two complete sentences, what you saw in all the cases you tried. Describe the pattern you experienced in general terms without using any actual number examples ... Let's write on the board some mathematical terms that may be helpful in your sentences: 'three consecutive numbers', 'square of...', 'product of ...' "

Example: The student has just experienced these:

5, 6, 7 ➜ 5 x 7 = 35, but 6 x 6 =36
3, 4, 5 ➜ 3 x 5 = 15, but 4 x 4 = 16
9, 10, 11 ➜ 9 x 11 = 99, but 10 x 10 =100.

["For any three consecutive numbers, the product of the first and last is one less than the square of the middle."]

➤ Following the solving of a story problem or the use of any calculation procedure, ask for a **general description of steps involved** to get the solution.

Teacher: "Describe, step by step, in general terms, what procedure you just used to solve the problem (or calculate the answer). Don't say any numbers as you write (or speak). Use complete sentences. After you have given the general principle, show how you are using it in the specific problem.

Example: "To get the total cost of all the computer disks, I first needed to find out how many disks there were. Then I could multiply that by the cost per disk. I only knew the number of boxes of disks that each have was 12. I multiply 5 boxes times 12 per box, then multiply this result by $0.88. I get $52.80."

[Such a write up is a good small group (interpersonal) activity, resulting in one optimized formulation from the whole group.]

Linguistically, mathematics can be taught as communication: as an extremely concise portrayal of relationships among quantities.

➤ Following the solution of a puzzle or problem, ask an **intrapersonal and linguistic task** related to the process.

Teacher: "Write at least two sentences about the AHA! experience – how you felt just as you glimpsed the way to solve the problem."

➤ **Give numbers drama and feelings.** Personalize students' perceptions through **poetry and prose**:

Teacher: "Write a short piece or poem about how 9 feels compared to 10, who is the 'big shot' of our number system."

Example: "I wish I could
            feel the fun
                of gaining just
                    an extra one."

➤ If a group gets stuck while solving a problem, students are asked to **formulate a question as a group** before asking for help.

Teacher (to working groups): "Don't ask me a question unless all in your group have the *same* question. Try to state your question about the problem in a complete sentence. Everyone in your group can help make the question. Then I'll answer it."

➤ Following the explanation of a concept have students **repeat the explanation.**

Teacher: "Repeat to your partner the explanation that I have just given. Let your partner add to it as necessary."

➤ Using a thoughtful or challenging problem, students **narrate or monolog their process** aloud.

Teacher: "Connect your brain to a speaker. That is, explain to a partner (called the "witness") what you are doing as you proceed to solve the problem. You don't need to smoothe out, revise, and make clear each thing you're thinking, just let it play on the 'speaker' which is your vocal cords.

"The witness' role is just to listen as you talk your thinking aloud, except, if you say something that the witness can't understand, the witness can ask you to clarify it. If you freeze up, the witness gently encourages you by saying, 'What's going on now?' Connecting your brain to a speaker gives you a chance to get clearer about how your mind works when you're solving a problem."

➤ Upon assigning explanatory material to read from a math text page, have students **summarize three main ideas.**

Teacher: "Read the page that explains about ... [a concept]. Write, in complete sentences, three main ideas that were presented about...[the concept]."

➤ After explaining a general concept, ask for a **discussion of a specific case** of it:

Teacher: "Can you, Sarah, say (or write) in a complete sentence what is really meant by ... [specific mathematical statement]?"

Example: " What is meant by 12 x 5 = 60?" ["It means 5 is taken 12 times and the total is 60." or "This means to take 12 groups of 5, and all these make 60."]

➤ After students have written information requested in the three previous suggestions above, have them **revise** it **with peer input.**

Teacher: "Read your partner's description of solving the problem (or his explanation or his summary), then give feedback about it. Compliment something about it and give at least one constructive suggestion. Was it clear? Was it complete? Did it use complete sentences? Was it accurate?"

➤ While studying any topic, have students **make a list of terms** and a **puzzle** that uses them.

Teacher: "Make a list of words that relate to the topic we are studying. Develop a crossword puzzle that uses these words for $^3/_4$ of the puzzle and any words you like for the other $^1/_4$ of it." (Best for cooperative groups)

➤ Help students develop **lists of conceptual ideas** based on a question.

Teacher: "I'm going to ask you a question. Let's brainstorm and list all the possible answers to it." [Discussion or written response. This can be added to as more ideas pop up during the year about the language connection.]

Example: "How is math like a language?" [It has sentences, like 2 + 3 > 4; it has symbols (like letters) that make "words" like 3578; it has "pronouns" like x that stand for "nouns" like 19; it has correct grammatical order, as in the sentence 23 + 7 x 5 = 58, where the multiplication must be done first because of the order of operations.]

➤ Consider **word origins** and analyze **word parts** as terms are introduced:

Teacher: "Before we go into this topic, here is the origin of the word, which will help us understand what it means."

Examples: Fraction comes from Latin *fractare*, to break. Multiplication comes from Latin *multi*, many, so multiplication means to "many-ize" a number.

Subtract comes from *sub*, under, and *tractare*, to pull; it means "to pull (a number) below or down."

Seasoning with the Spatial Intelligence

This intelligence is often called visual-spatial because the visual is usually the channel for activating and processing the spatial intelligence. (With blind persons the channel to spatial intelligence is usually touch.) About 40% of children are visual learners, thus elevating the spatial intelligence to a prime position in communicating and understanding mathematics. A visual activity may be *highly* spatial or, as in symbol manipulations, *barely* spatial. *(See the Resource Bibliography: Seymour, Taylor, Wahl, and others, for more visual and spatial math activities.)*

Here are several ways to bring more of the spatial intelligence into your daily teaching:

➤ When explaining a numerical procedure (algorithm), **make a diagram:**

Teacher: "Here is a diagram that shows the steps you do to get an answer."

Example: To convert an improper fraction to a mixed number, the upward arrow conveys division of the denominator into the numerator; the lower of the split top arrows indicates the whole number answer to the division; the upper of the split arrows goes to the remainder. The bottommost arrow shows the same denominator being kept. See all movements along the arrows as a graceful, linked motion of the calculation.

We inherited geometry from the ancient Greeks. Geometric figures are a spatial way to represent and symbolize numerical relationships.

➤ When studying a series of steps in a procedure, make a **flow chart.**

Teacher: "Make a flow chart that shows the steps you must do to ... [name procedure]."

Example: To show steps for adding two fractions:

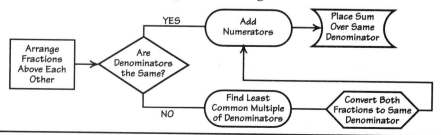

➤ Use **color** to distinguish the parts of a calculation or figure.

Teacher: "Write the numbers above each other with the ones digits red, the tens digits blue, and the hundreds digits green. When you regroup and carry a number to the next column, write the carried number in the color for that column."

➤ Use **manipulatives** to model math concepts and skills. Manipulatives have both a visual/spatial component and a tactile/kinesthetic component. *(See "Seasoning With The Bodily-Kinesthetic Intelligence" for more discussion of manipulatives.)*

➤ Encourage students to make a **picture** or **design** with something they are trying to remember.

Teacher: "To help you remember ... [points to word or configuration], make a picture that uses ... [that word or configuration] as part of it."

Examples: "Make a picture out of these symbols: 8 x 7 = 56."

"Make a picture that has the words 'Average', 'Add' and '÷2' in it. The words can form part of the picture. This will help you remember that to find the average, first add, then divide by two (for two numbers)."

➤ To help students remember a term or manipulation, use a **visual-spatial mnemonic**.

Teacher: "We're going to make an image to help us remember this concept."

Examples: "The hypotenuse of a right triangle is as big as a *hippopotamus* – it's the biggest side. [Draw a hippo as the longest side of the triangle.]

"On a screen inside our minds we're going to see *many* of a certain kind of *product* (like boxes of cereal, bottles of detergent) coming off an assembly line. This will remind us that the result of multi-(many)-plication is called a *product*."

"8 x 7 = 56 forms a triangle, so we can see that there are *two* division facts that go with any multiplication fact."

➤ Include **geometry, geometric design,** and **measurement** as much as possible in your instruction and enrichment activities. *(See the Resource Bibliography: Seymour, Taylor, Wahl.)* It will be highly motivational to the spatially-intelligent students who may have tuned out on symbol work.

Examples (Practicing division and multiplication facts): "Draw a rectangle and write 54 sq. in. inside it to mean its area. Its length is three more than its width. What are their values?"

"This week's extra project is a mandala using arcs and straight segments. You will need a protractor to mark off equally-spaced points on a circle. Confer with members of your group about ideas for making it elaborate and striking. Color it completely, and we will create a bulletin board display in the hall on Friday."

➤ Review concepts with a **mind map** or **chart**. *(See the Resource Bibliography: Margulies.)*
Example:

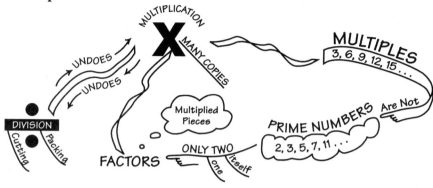

➤ Use **videos, slides, posters,** and **computer software** that represent math concepts visually.

➤ Relate concepts to **graphs, Venn diagrams, branching trees.**
Example: "If a number greater than 2 is prime then it is odd. But if a number, like 15, is odd, it is not necessarily prime.

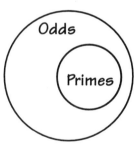

"If there are three colors (Red, blue, and Green), three sizes (S, M, L), and two styles, A and B, how many different items are possible?"

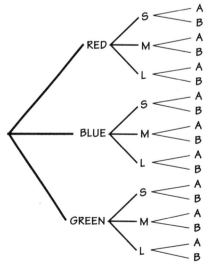

Internal visual imagery is a powerful spatial exercise. It may be challenging for you at first to guide students to explore math with it. After a few trials they will beg you for more!

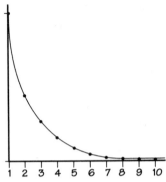

"This graph shows how the size of a fraction decreases as the denominator increases."

➤ Use **internal imagery** or guided imagery to relate to concepts. *(See the Resource Bibliography: Gordon, Mason.)*

Teacher: "Close your eyes,... , breathe deeply,... , relax, ... See three bright orange dots on the screen of your mind. Join them with line segments to form a triangle. Let the dots slowly move around while still joined. What different triangles do you see? ... Slowly open your eyes and begin to draw what you saw."

➤ As a memory device have students write formulae, words, or equations *upside* **down** (so someone facing them can read it right side up), or in **unusually-shaped symbols**. This activates the right brain and spatial intelligence.

Teacher: "Try to write 8 x 7 = 56 upside down. Then write it with extremely tall, thin characters."

Seasoning with the Musical-Rhythmic Intelligence

A vast amount of data supports the fact that music influences learning. It can have its effects preceding study, accompanying work, or by actually being the subject of study, like fractional notes and rhythmic patterns.

A surprising piece of data is that Mozart is among the top math enhancers! Ten minutes of the "Sonata for Two Pianos in D Major" has been shown to measurably enhance reasoning, bolster memory, and strengthen spatial thinking. And, even better than the effects of silence or relaxation tapes, Mozart has raised scores on tests that require spatial and verbal reasoning *(See the Resource Bibliography: Jensen, Mozart).*

Here are some ways to sprinkle Musical Intelligence into your math lessons:

➤ Play 3-10 minutes of **"conducive" music before undertaking learning tasks.** [Start with Mozart then try some other classical, baroque, and romantic composers as your confidence grows.] *(See the Resource Bibliography: Mozart, D. Campbell.)*

Try teaching the actual math of music theory, from fractional notes to the frequencies of tones.

➤ Use **soundbreaks**: music between activities to pick up tired energy and low moods (e.g., the theme from "Flashdance"), activate imagination (Paul Winter's "Earthbeat"), or focus intent (Bach's "Well-Tempered Klavier"). Use music sparingly, perhaps twenty minutes, during certain tasks requiring repetitive work tasks (Baroque) or group interaction (softer New Age). *(See the Resource Bibliography: D. Campbell.)*

➤ **Music theory**: Teach fractions from notes, *hear* number patterns (see Activity page 5-6. *Also see S. Beall, Resource Bibliography*).

➤ Use **clapping, singing, humming, rhythmic movement, rhythmic words (raps), or jingles** to help students recall and recite concepts, tables and procedures.
   Examples: [Rhythmically] "A hun … dredth … is a TENTH of a TENTH … (I said) A hun … dredth … is a TENTH of a TENTH …" [whole class snapping fingers, saying a few times] then start, "thousENTHS … are TEN times smaller … (I said) thousENTHS … are TEN times smaller…". [Teacher points simultaneously to each place mentioned in a decimal number on the board.]
   Sing the fours table to "Jingle Bells" (starting with "Dashing through the snow…"): "One times 4 is 4, (and a) two times four is 8, 3 times 4 is 12, four 4s are sixTEEN, … five 4s are twenTY, …, etc."

➤ Ask groups of students to **compose songs or raps** that capture or abbreviate a particular concept.
   Examples: [Snap fingers in rhythm] "When the bottom grows, the fraction slows." (Used to recall that larger denominators make a fraction smaller.)

➤ Use **musical language and metaphors** to describe math ideas.
   Examples: "The thousands, millions, billions, and beyond, are higher 'octaves' of the one, ten, hundred idea. That is, there's a one, ten, hundred in the thousands, another in the millions and yet another in the billions. These are like the same notes, but in higher octaves." [Teacher demonstrates on an instrument, or 'sings' with the class a number with high voice '652 million,' medium voice '586 thousand,' and low voice '297.']

"All the even numbers dance to two's tune."
"A prime number, like 7, is a pure tone."
[Teacher makes it on an instrument.]
"So is 3." [Makes another tone.]
"Then 21 (=3 x 7) is a composite 'chord' number." [Plays both together.]

Seasoning with the
Bodily-Kinesthetic Intelligence

Many students that get in trouble in the school setting are kinesthetic. They need movement, action, and physical contact. When most teachers don't provide it, they create it by falling off chairs, horseplaying with classmates, throwing things, chewing things, doodling, and crumpling unused paper.

Without turning the class into a three-ring circus, there are many ways to feed the kinesthetic intelligence in math students.

A major kinesthetic tool for the math instructor (that feeds students' spatial intelligence as well) is the use of **manipulatives** or hands-on models. There are numerous manipulatives available (enough to bankrupt your materials budget!), so I will recommend what I consider the five most useful. *(For more information on all of these and how to obtain them, consult the Resource Bibliography: Manipulatives).*

Omni-directionally Linking Cubes. These should be of at least four colors and able to stand alone or link in columns, planes, and solids. They should come with a good manual and be priced reasonably enough to purchase at least 600 for a class. These cubes can be used to create good models for concepts of place value, regrouping in addition and subtraction, simple multiplication (two-digit by one digit) and division (by one digit). They can also model fractions and some geometric ideas.

Base-Ten Blocks. After you use simple bean counting and 10-bean popsicle sticks, the base-ten blocks present the next level of sophistication in place value, because they can spatially model up to the thousands place. After this, the cubes mentioned above take over to represent tens, hundreds, and thousands as just different colored cubes. That is, three reds might be just 3, but three blues would be 30, and three greens would be 300, etc. Trading with all of these can simulate regrouping, i.e., "carrying" in addition, and "borrowing" in subtraction. Besides this, the ones, tens, and hundreds pieces of the base-ten blocks make good beginning models for decimals (up to hundredths). Some products extend them to thousandths (using tenths of a one-cube).

Cuisenaire® Rods. Among the old standards that still prove useful are Cuisenaire rods. They model many things, but I like them especially for addition facts and multiplication ideas.. They are in exact-centimeter lengths, making the meter stick an excellent adjunct tool. For instance, eight blue 9-rods laid end to end will just reach the 72 mark on the meter stick. The student can directly experience that $8 \times 9 = 72$ this way. In showing 7×13, a student will instinctively lay seven 3s and seven 10s on the meter stick, which exactly parallels the pencil-and-paper algorithm for multiplication.

Fraction Pieces, Bars, Tiles. If you are low budget, mark multi-colored small dessert plates using a protractor then have your class cut them in halves, thirds, quarters, sixths, eighths, ninths, and twelfths (with fifths and tenths optional). Similarly, fraction strips of these various fraction sizes ruled on unit strips can be made from colored paper or acetate, and you can purchase manuals on how to use them. Most fraction concepts from equivalency to division can be seen and felt through the use of these — a must if you want to get away from the blank looks and eternal re-teaching of fraction concepts.

Pattern Blocks. This manipulative creates spatial, geometrical learning. It encourages exploration. The blocks can be used to create elaborate symmetrical or asymmetrical geometrical designs that depend on the interaction between geometric shapes. They can also be used to model simple fractions in unusual forms.

Calculator. Because the calculator gives a tactile feel to math and involves fine motor motions, I include it in the list of important manipulatives, but for many reasons it should be an integral part of any math program. Any modern textbook and numerous publications show how to make it so. *(See the Resource Bibliography: Schielack, Williams)*

Other useful manipulatives. These can be added, at your discretion, depending on the levels you teach, after you have mastered the basic ones. *(See the catalogs at the beginning of the Resource Bibliography.)* They are:
- clock face with moveable hands (have your kids make them from paper plates)
- 3-D shapes: cone, pyramid, icosahedron, prism, etc.
- compass for drawing circles
- geoboards for modeling geometric relationships (can be homemade)
- scales that weigh in oz., lb., gm, kg
- 1-100 number grid for exploring patterns (can be homemade)
- containers: gal., qt., pt., cup, fl. oz., L, 100 ml
- $\frac{1}{2}$″ plastic tokens in at least four colors for place value work (chip trading) and placing on a number grid
- ten-sided dice for number games
- attribute blocks for teaching logic

➤ Use **kinesthetic (action) as well as tactile metaphors and language** for math concepts.

Examples: "The denominator of a fraction cuts (saws, shreds) a whole thing into pieces. [Pretend you are a denominator cutting your desk into five pieces.] The numerator *chooses*

(picks) a number of them. [Pretend you are picking three with a pointing finger.] You have just done a $^3/_5$ to your desk."

"*Area* means the amount of *smooooothe* flatness inside a figure. Rub your hand on the (large) figure's inside. Perimeter has *rim* hidden in it. It's the length of just the rim. Run your finger along the whole rim."

(Note the physical motions and meanings of the operations in Chapter 9: "+ - x ÷ Supercharged!")

➤ Ask students to simulate math ideas with their **whole bodies.**
Examples: "Nine students will bend and twist to be the nine digits of 891,473,265. Two more students will be commas. Each student I point to should say what they mean [e.g., the 7 student means 70,000]."

"Each group of four students will somehow make a parallelogram with your bodies. Keeping both pairs of sides parallel, change the shape a few times."

"How could we quickly find the median height of this class without pencils, paper, or calculators and only a short ruler?" [Kids figure out they can line up in order of height then find the middle person and measure, or find the in-between height of the two middle persons, if there is an even number in the class.]

➤ Play **physical games** to drill math concepts.
Examples:
Hopscotch or Twister redesigned for math facts practice.
Jump-rope for rhythmic or rhyming repetition of concepts.
Card games, board games, like concentration, Uno®, rummy, bingo, fish, Scrabble®, and Trivial Pursuit® redesigned for use with number cards and markers.
Students can help with the design and rules of the game.

➤ Have students **role-play action problems** before solving.
Example: "I am a scout and must locate hidden ammunition in a field. I can search 80 square feet every three minutes. The field is 160 feet by 120 feet. My partner must search the perimeter fence for hidden marks. She can check four feet every five minutes. Who will finish first and by how many minutes?"

First a few students interpret the problem and act out the search while indicating what the numbers in the problem refer to — on a small classroom area or outdoors. Then in groups, some compute the search time for one scout and some for the other.

The computer industry has proven that the more visual and active they make abstract processes, the happier their customers become. This is true for our students, too.

➤ Preceed math lessons with **physical "tuning" exercises** that multiply learning efficiency. It is well documented that physically stimulating and coordinating movements carry over to mental potency. *(See the Resource Bibliography: Hannaford, et al.)*

➤ Request that students **make products** using their math skills. **Examples:** Create 3-D geometrical models and 2-D geometrical designs. *(See the Resource Bibliography: Seymour, Taylor, Wahl.)*

Make a scale model of our planetary system with the sun the size of a large inflatable beach ball or an imaginary sphere that is the height and location of a room in the building. Let students research, compute, and build in groups.

➤ Work with **real life situations** — they involve more bodily and tactile experience.

Examples: "Figure out your parents' credit card bill, phone bill, or power bill *[for the power bill, see the "Watt's Happening?" Activity]*, check all the numbers on it, and explain it to the class."

"Go to a car dealer and find out all the variations that cause price changes. Compute the cost of several car 'packages.'"

Go on **math field trips** (e.g., engineering firm, surveying crew, insurance actuarial office). They involve bodily and tactile experience, with spatial and interpersonal skills included. Be sure to plan learning activities related to the trip content *before, during,* and *after* the actual trip so as to harvest the most from the direct experiences and data encountered.

➤ Make physical **puzzles or board games** that review and connect information.

Example:

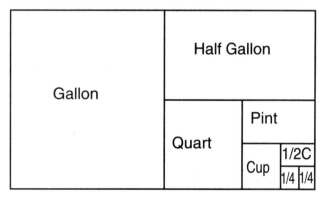

Cut this into pieces and reassemble

Seasoning with the
Intrapersonal Intelligence

The intrapersonal intelligence is often ignored in classrooms; math has a reputation for being cold, calculating, and objective—no room for anything personal. But, an integral part of mathematics is the feeling, thinking student. Acknowledging and discussing the wide range of feelings triggered by a math lesson can better link a whole class to the material. "Tracking" a student's math thinking process aloud can illuminate strategies for the whole class. "Metacognitive training" teaches marginal students how to identify, and self-monitor progress towards, personal learning goals. All of these lead to greater student autonomy and higher grades. So, begin seasoning in these ways:

➤ **Metacognitive training:** Guide students to self-identify learning and grade goals, then self-monitor progress and course-correct.

➤ **"Interview"** students to closely track and highlight internal thought and feeling processes: a benefit to them and the whole class.
  Teacher: "Let's trace out loud exactly what thought (and feeling) steps you experienced from the moment you first saw the story problem until you successfully got the answer."

➤ Pay attention to, work with, and try to alleviate **math anxiety**. (*See Chapter 7: "Dealing with Math Anxiety."*)

➤ **Share your internal process,** where appropriate, with the students. Be **self-revealing** rather than remotely authoritarian.
  Example: "For some reason I'm stuck on this problem myself. What I find helpful here is to leave it for a while so it can incubate, then the solution sometimes pops into my head. I'm going to take it home tonight, and I'm sure I'll figure it out. I'll bring back the results tomorrow."

➤ Notice, acknowledge, and allude to **feelings arising in class**.
  Examples: "Mary, I heard you say , "Wow!" Was that because you enjoyed the pattern the answers made or because you were successful?... What's it like to enjoy a math idea when you have usually thought math was not for you?"
  (Privately:) "James, I just saw you roll your eyes when I asked everyone to average the numbers. Does that mean you are bored with it, or do you feel unable to do it? I'd be happy to get you started if that'll help."
  "I can feel frustration and fatigue arising in class as you try to learn this confusing concept. Let's do something completely different right now and I'll try to present the concept differently tomorrow."

Students read you intrapersonally. If they can feel your passion and enthusiasm, it will ultimately be your most successful teaching "technique".

➤ Attribute **feelings or personality** to mathematical entities. **Examples:** "Imagine how a **1** feels in Multiplication Land: it's a nobody. Everyone it multiplies doesn't feel *anything*. They just shrug and walk away exactly the same. **1 x 5** just stays 5. But at least **1** can take a vacation and go to Addition Land where it can make numbers change a little. Now discuss how **0** feels in both lands."

"The denominator 5 of fraction 4/5 is a 'cutter.' It loves to cut things into 5 pieces and let the numerator 4, the 'chooser,' pick 4 pieces. But a cutter gets very upset when a chooser/ numerator demands more than it cut (as in the *very improper* 17/5). Then it must go find two more wholes to cut in 5."

Seasoning with the Interpersonal Intelligence

This intelligence is extremely valuable in learning mathematics. There are several excellent ways to sprinkle it into your lessons. *(Some verbal-interactive ways are found in the Linguistic Intelligence section of this Chapter.)*

➤ Use **cooperative learning** techniques regularly. They will multiply the effectiveness of most learning experiences. Turn your class into a trusting, interactive environment. *(For details, see Chapter 6: "Cooperative Learning.")*

➤ Allude to the **cross-cultural and historical** aspects of the topics the class is studying. *(Found in many "History of Mathematics" titles in any library. Also see Resource Bibliography: Knauff.)* **Examples:** "There have been many different symbols for numbers used throughout the world through time. Today, the number symbols we are using are the most universal language in the world. You could write a sentence in these symbols, like 87 x 3 = 261, and it would be understood in 99% of the populated places of the earth. These symbols were not invented by Europeans or Americans. Does anyone know where they came from and how old they are?" [They're originally from India, invented around 600 AD, and they were spread from there through the Middle East and northern Africa by the Arabic Moslems, who conquered India. The numbers slowly made it to Europe by 1200, AD finally used extensively there by 1450 AD, instead of Roman numerals.]

"When was a decimal point first used?" [Probably in 1592 in Italy by G. A. Magnini, a map-maker. Decimals were made into a standard notation by John Napier in 1617.]

➤ Encourage students to play two- or three-person **math games** to practice skills. [Most modern math books and their teacher manuals contain many such games; also see Activities 2 and 4.]

Seasoning with the Logical-Mathematical Intelligence

You're *teaching* mathematics, so why *season* with this intelligence too? In many math classes, the logical-mathematical intelligence might be exercised very little in the process of copying down information and memorizing meaningless maneuvers. *(See Chapter 2: "The NCTM Standards," to get an idea of how many more mathematical elements can be included.)*

The key here is to make sure your lesson works the logical reasoning and mathematical capacities of your students. This usually means that you need to bring them beyond a level of *knowing* a set of facts, and even *comprehending* these facts, to *application* and *analysis* of them. Here are some ways:

➤ Work a concept **backwards** after they have learned it forwards.
Examples: "Now that you have learned to multiply two-digit times two-digit numbers, can you sleuth the missing digits that will make this a valid calculation?"

$$
\begin{array}{r}
3\ \square \\
\times\ \square\ 6 \\
\hline
1\ \square\ \square \\
\square\ \square\ \square \\
\hline
1\ \square\ 9\ 2
\end{array}
$$

"You have learned that the area of a triangle is $^1/_2$ of the base times the height. What would be the *height* of a triangle whose area is 60 and its base is 12?"

"You have learned to multiply two fractions. What fraction would $^2/_5$ multiply by to make $^{14}/_{25}$?"

➤ Turn single-step procedures into **multi-step procedures** with **story problems.**
Examples: "You have learned to add and subtract and multiply. Now, instead of practicing each separately, try to solve this problem that's on the overhead: 'For Christmas gifts you have purchased three Lego kits for $3.12 each, four Frisbees for $4.69 each, and three CDs for $7.96 each. Your sister bought five bike lights for $4.26 each, two CDs for $10.95 each, and three earmuffs for $3.89 each. Each is arguing that they spent the most. Who did, and by how much?'"

Following the learning of multiplication of fractions: "Tessa spent five hours gathering aluminum cans at the county fair to recycle, and her friend Alex spent three. Together they got $42.56 for the aluminum. How should they

fairly (pardon the pun) divide it?" [This calls for division by 8 then multiplication by 3 to get Alex, or it calls for assigning $3/8$ to Alex and $5/8$ to Tessa, then multiplying $3/8 \times \$42.56$.]

➤ Teach **inductively** rather than deductively. That is, encourage students to explore situations then **conclude a general rule** that governs them, rather than the reverse — giving a rule then asking students to use it in examples. (*Most material in books is introduced deductively; turn the same lesson into an inductive one.*)

Examples: "Here are six answers we got when we multiplied numbers by 11 on the calculator:

$23 \times 11 = 253$ $52 \times 11 = 572$ $34 \times 11 = 374$
$31 \times 11 = 341$ $62 \times 11 = 682$ $81 \times 11 = 891$

Can you make a general rule that predicts an answer when you multiply by 11?" [For numbers with digits that sum to less than 10, re-use the first and last digits and put the sum of these digits in the middle. The rule can be adjusted for numbers whose first and last digits sum to more than ten. Try some examples.]

"Here are five illustrations of right triangles. Can you write the definition of a right triangle?"

"Here are three calculations I have done, each for division of a fraction by another fraction. Can you figure out from them a general rule for how to divide two fractions?"

➤ Pose **thoughtful questions about ordinary ideas.**
Examples: "Why do you think a dozen is used as a standard amount and something like 9 isn't?"

"What do you think was the first numerical idea thought by a human being?"

➤ **Provoke exploration** rather than just telling.
Examples: "Now that you know how to multiply two *proper fractions*, see if you can figure out how we could multiply two *mixed numbers to give a meaningful answer.*"

"Using your ruler, draw the classroom so that everything is exactly the right miniature size. That is, make it look proportional to the big classroom."

"Are there more multiples of 3 or multiples of 8 among all the counting numbers?"

"Is it true that larger numbers have more factors?"

➤ **Create disequilibrium** and **invite class discussion** of concepts through meaningful, surprising questioning. Encourage clear, logical argumentation in presenting the answers.
Examples: "I think $0 \div 0 = 0$. I'll check it: Sure enough, answer 0 times divisor 0 gives 0! It checks! But shouldn't the

answer be 1, since dividend and divisor are equal? I'll check this answer: 0 x 1 = 0! It checks! But maybe the answer is 17! It checks! What answer couldn't it be? [It could be any answer.] What should mathematicians say the answer is?" [They say 0 ÷ 0 is a meaningless question; it has too many answers.]

"Does .9999... exactly equal anything? You've learned that $1/3$ = .3333... and $2/3$ = .6666..., so add them digit by digit: .3333... + .6666... = .9999... But $1/3 + 2/3$ = 1, so it must be true that .9999... = 1. I mean <u>exactly</u> equals, not approximately, or rounded off. Discuss the validity of this argument in your groups."

"What is the biggest number less than three? Justify your answer." [There is none since you can always find a closer and closer fraction or decimal to 3. Kind of like Aesop's fable where the dog won't catch the rabbit (3) because he has to go through an infinite series of getting to half his previous distance from the rabbit (3).]

➤ Encourage **mental math daily** (*See the Resource Bibliography: NCTM*). Whenever possible, request that the students attempt to do the simpler calculations **in her or his head**. This may involve thinking of the numbers in a very different way than just reproducing paper and pencil in the mind.

Examples: 86 - 19 is best done not by borrowing, etc., but by noting that 19 is like 20. So 86 - 19 starts 86 - 20 = 66. Then add 1; 66 + 1 = 67.

12 x 25 can be done simply if we appreciate that four 25s make 100. Twelve has three fours of 25, so 12 x 25 = 300. Such methods utilize the right brain perspective and spatial — even kinesthetic — intelligence that sees (feels) patterns and interrelationships in numbers. Less useful here is the sequential left brain function that relies on fixed digit rules and isolated manipulations. (*See Chapter 4 on "Brain Hemispherity".*)

➤ Encourage **estimation** daily. Often preface a request for a calculation with the question **"About how much do you expect the answer to be?"** This activates other intelligences also, like spatial and kinesthetic, as well as right-brain wholistic functions that look for contexts, relationships, spatial sizes, etc.

Examples: "How much so you expect the answer to 24 x 18 to be?" [25 x 20 = 500 would serve as a high approximation. This isn't exactly rounding, just moving to numbers that are easy to work with, like 25.]

"About how much will $7/15$ of 88 be?" [Since $7/15$ is just under $1/2$, we predict that the answer will be less than 44, maybe 40 (it's 41).]

"Before you measure, estimate the length of the rectangle."

➤ Ask students to **make up their own story problems** that require a given computation. (Remember, a problem must state facts that have numbers in them, then it must conclude with a "How many?" or "How much?" question.)

Examples: "$456 \div 24 = 19$. Give a real life problem involving school lunch where this computation would be needed."

"Here is a computation: $\frac{3}{4} \times \frac{2}{3} = \frac{1}{2}$. Make up a real life problem about horses and hay in a stable where this computation would give the answer." [A sophisticated answer: "A stable worker had $\frac{2}{3}$ of a bale of hay left and placed it near a horse stall noting how far it reached along the wall. Two days later, the worker observed that $\frac{3}{4}$ of this hay was eaten. How much should she report that the horse ate in those two days? ($\frac{1}{2}$ bale.)]

➤ **Make connections** with other areas of math as you teach a concept.

Examples: Something as seemingly simple as the fraction $\frac{1}{12}$ is a jumping-off point in many directions. Any one of the starters mentioned here can lead to the others.

Rather than just seeing $\frac{1}{12}$ as written ciphers, it can be seen on a clock face as the angle between two numbers, thus representing five minutes. Geometrically speaking, $\frac{1}{12}$ of a circle is 30° (so this is the angle between any two numbers on a clock face).

Graphically speaking, it, like all fractions, can form a uniquely sloped line by starting at the origin, going over 12 and up 1 then making a dot. In algebra, as in carpentry, $\frac{1}{12}$ is called the *slope* of the line or of the roof. Also in pre-algebra, $\frac{1}{12}$ is the solution to $12\,a = 1$. In probability, $\frac{1}{12}$ is about the chance that a randomly chosen person walking by on the street was born in November. Computationally, $\frac{1}{12}$ of a number is found by dividing that number by 12. Also $\frac{1}{12}$ is equal to a decimal (.08333...) and a percent (8.333... %), each expressing one of 12 parts of 1 or of 100. In measurement, an inch is $\frac{1}{12}$ of a foot.

Historically, the Chinese, Egyptians, Greeks, and Hindus all had different systems for symbolizing $\frac{1}{12}$. This is a good research project.

➤ Encourage **problem-solving thinking daily**. [Many suggestions above already encourage this. In addition, utilize real life situations, puzzle problems, and exploratory problems often. Develop a classroom list of strategies that have helped students solve tricky problems. For a multiple-intelligences example of such a list, see "Get Really Smart: Crack Problems with All Your Intelligences" in the Activities section. Also, many of the other Activities model problem-solving thinking.]

Don't settle for imprecise comprehension of mathematical processes. Try to make your, and their, math language exact as you convey each concept.

Seasoning with the Naturalist Intelligence

➤ Use **natural settings** to instruct math.

A student with strong naturalist intelligence (who has access to natural locations in her environment) will generally respond well to learning math in a natural setting, even if she is normally not well-motivated in math and even if the math doesn't directly deal with the setting. Even more learning and motivation results, of course, from using the setting as part of the math lesson.

Example: "Estimate the number of needles on this evergreen tree. Back up your estimate with counts and calculations that justify your number."

➤ Use **natural objects** to **model** math concepts and processes.
Examples: Use petals, small buds, hard berries, or small rocks as counters to model sorting, counting, subtraction, place value (see bean-sticks discussion p.8), averaging groups, etc.

Explore the Fibonacci Number patterns in daisies, sunflowers, pine cones, pineapples, bee colonies, rabbits, etc. (*For instructing this see the Resource Bibliography: Wahl.*)

Use family trees of animals or people to model fractions (i.e., show why if only one grandparent is pure German and the rest are Italian, then the grandchild is $1/4$ German, etc.)

➤ Use **sorting** and **classifying** as a way to make math terms and processes comprehensible.
Examples: Use a "family tree" or Venn diagrams to show how whole numbers and negative integers are members of the integers, integers and fractions are members of the rational numbers, then rationals and irrationals are members of the real numbers.

Do a "quadrilateral family tree" relating squares, rectangles, parallelograms, rhombi, trapezoids, and kites.

"Program" the "Number Detector" to detect characteristics of several numbers above 16. (Activity pages 1-1 to 1-5)

➤ Study **natural processes** or **objects** outdoors with math tools first developed in the classroom. This will motivate the study.
Example: Develop the concept of finding the volume of a rock by placing it in a full beaker of water and measuring the water overflow with a graduated measuring cup. Then ask students to find a group of rocks outdoors and line these up in order of their volume size with volume labels (fractions of a cup) on them.

Brain Hemisphericity

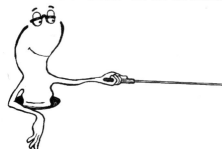

BOTH HALVES HELP MATH LEARNING

There's been a lot of talk in the last two decades about the fact that we really have two brains in our heads, not one. We possess both a "sequential" brain (usually the left half) and a "wholistic" brain (usually the right half). These two brains have very different approaches to processing and comprehending information, yet both are housed in the same skull.

Fortunately, your half-brains have a "cable" and "switchboard" (consisting of some 300,000,000 nerve fibers) between them, called the *corpus collossum*. This connection insures that there will be some coordination between the strengths of both sides of your brain. In order for you to understand where some kids' difficulties in math arise and to expand your repertory of approaches to teaching, it's helpful to know about the "hemisphericity" of your and their brains. Most lean toward dominance of one or the other side. I'll give a kind of "pop" description of the phenomenon — just enough so you'll get the understanding and images you need, and to get you on the road to learning to switch your — and your students' — brain functions at will!

This description is by no means intended to be detailed enough to please a brain researcher, though the generalities that follow are basically accurate. For our purposes, the notions of left (sequential) and right (wholistic) brain are metaphors for two fundamentally different approaches to life and learning residing in each person's brain.

I once heard the Nobel nominee, Robert Ornstein, describe some surprising research, which drove home to me how different the two sides of the brain are. He and Roger Sperry devised some subtle experiments. They knew that the right brain receives input from the left half of each retina, and the left brain receives from the right halves of the retinas. The left half drives the right hand and the right half drives the left hand. With these in mind, small blinders were rigged to each eye lens of a subject so that either half could be closed off, thus allowing images into a selected side of the retinas. That's the *input* to each side of the brain.

Output from each side of the brain could be obtained by having the subject write or draw with the appropriate hand. However, there was one critical factor: there must be no "bleedthrough" of thought from the other side of the brain through the corpus collossum. The key was to use subjects that had a *severed* corpus collossum! There just "happened" to be some around — severely afflicted epileptics whose corpus callossa had been severed surgically, since that procedure seems to create some relief from seizures.

The chosen subjects led fairly normal lives since both sides of their brains generally got bilateral input from their senses. But in the carefully-controlled experiment a very interesting distinction emerged. A picture of this cube (see fig. 6.a) was shown to the right brain which seeks wholes, relationships between things, contexts, and spatial relationships. The subject drew with his left hand what his right brain had seen (see fig. 6.b); nothing too surprising.

Now the same shape was shown to the left brain, then the right hand was asked to draw what was seen. Here's what it drew (see fig. 6.c):

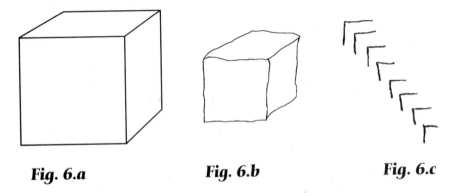

| **Fig. 6.a** | **Fig. 6.b** | **Fig. 6.c** |

What's going on here? The left brain got caught in some details (i.e., the eight corners of the cube). To the left brain, a cube was eight corners arranged somehow, but the *relationship* was elusive. The arrangement was of little consequence. The <u>fact</u> was important, but the <u>context</u> of the fact wasn't.

So, for our purposes here, we will consider the strengths of the left and right brains in a schematic way. This is an oversimplification of the true brain physiology, in which some of these functions have several subfunctions, a few of which may overlap two sides of the brain. But this will be sufficient to give us two major classifications of brain functions for discussing math processes. (*For more depth see Resorce Bibliography: Ornstein, Hannaford.*)

Abilities of the left brain

Sequential. This means that the left brain is more of a "one-track" mind. Often the word "linear" is used to describe it. A string of steps or logically developed ideas is preferable to an interwoven network of impressions. The linearly printed page is comfortable to the left brain, and so is our one-word-at-a-time language for explicitly presenting ideas.

In mathematics teaching this approach corresponds to learn-

Mathematics is best done utilizing materials and experiences that tap <u>both</u> sides of the brain.

ing sequences of procedures on digits (called "algorithms") on letters (called algebraic rules) to get answers. Doing these rules goes toward some answer, but can the rule know if it is the answer desired in a story problem? The calculating procedure can't make *tactical* choices, only *mechanical* ones. Without understanding the whole context (best done by the right brain), we can't really know if a mechanical procedure we are doing is the correct one.

Strongly left-brain-dominant teachers love to have the kids in straight rows, the grades in ruled lines in the gradebooks, and the day in a tidily-divided schedule of subjects. Such a person teaches math as a set of sequential rules about how to calculate something. Less important is why the rules work, their ultimate purposes, or their contexts.

Digital and symbolic. The left brain prefers handling numbers as a set of symbols to be manipulated by cut-and-dried rules. Every number looks roughly the same, though the digits can be seen to be different. "Feeling" or "sensing" the meaning or size or interrelationships of the numbers is irrelevant.

Detail-oriented. The left brain focuses on the trees rather than the forest. The details may not seem all that related; each fact is kept track of separately. Each piece of the puzzle is an independent world unto itself.

Analytic. Analyzing, for the left brain, means to tear a situation into smaller and smaller pieces to find out what's it's made of, then to try to fix that piece. (Current politics, medicine, and business, for example, often get stuck in this approach.)

Abilities of the Right Brain

Wholistic. This means that it perceives the whole, the context, the interrelationships, better than it perceives the details. It goes for the "gist" of an idea rather than the "nitty gritty." It looks for the layout of the forest but may ignore individual trees. The right brain wants to know what the subject in math is about, why it's being learned, what the tools do, and the like, in a general sense before connecting into the specifics. Then among the specifics it will look for patterns, interconnections, etc.

Spatial/visual. The right brain function seems to look for spatial and visual (even kinesthetic) cues to create context. It likes to see how a piece of information is imbedded in a whole. It tends to conceive of abstractions spatially — as, for instance, seeing a complicated love affair as a "triangle" or a political situation as a playing field. In mathematics, it needs to feel that 35 is in the third layer of tens or that an equation is a felt balance of two quantities, or that a fraction is part of a pie.

A clock with hands is a more spatial message about time's passing than is a digital watch, where number symbols alone are interpreted as a position in the day's timeline. Even color, inner visualization, daydreaming, and movement strongly activate the visual, kinesthetic, and spatial references in the right hemisphere. When was the last time you experienced these elements in your math learning?! Most teachers report that they rarely did.

Intuitive. Intuition is a complex affair that utilizes information from both brain hemispheres but seems to be led by the right brain. An intuitive person grasps subtle patterns and sees a single incident in a *context* rather than in isolation. Intuitive information often seems to come as a "flash" because a *pattern* has just emerged from scattered information. Intuition must be encouraged in most learners since former learning experiences may not have even acknowledged it.

The recognition of *pattern* seems to rely more on the right brain. For instance, these numbers show up in pine cones: 5, 8, 13, 21. The left brain says, "That's fine," and sees each number as separate. The right brain says, "The numbers are related. Each two add together to give the next. I can 'see' what the next one and the earlier ones in the pattern would be."

Tonal. The right brain picks up on the *tone* of information, the "between the lines" relationships rather than the content only. Thus it looks for the "gist" of the story and bypasses the details. And literally, musical melodies and rhythms activate the right brain while the words activate the left. Also, for some people numbers have tones and shades of color and feeling.

"Real Life" and "Common Sense." Sometimes these phrases evoke in us a shift to the right brain because, in coping with our daily lives, we are often estimating, feeling out things, filling in the missing information from a context, and the like. That is, we often rely on our right brains to shortcut us through the avalanche of detail that daily life brings. So, even though the left brain also deals with real life and can be a source of "common sense," those phrases tend to shift us out of a left brain fixation.

The right brain has a very real role in mathematics despite the extremely left-brained, digital approach to math training we've all experienced. The ideal for a mathematician is to draw on *both sides* of the brain.

Einstein certainly did this. When asked by an accomplished mathematician to describe the process of his mathematical discovery, Einstein answered, in a letter: "The words or the language, as they are written or spoken, do not seem to play any role in my mechanism of thought ... Before there is any connection with logical construction in words or other kinds of signs ... [my elements of thought] are of visual, and some of muscular, type. Conventional words or other signs have to be sought for laboriously only in a secondary stage ..."

Yes, intuition, movement, color, mystery, feeling, estimation, touch, and rhythm all have a place in math learning! But, in most math teaching today these are barely beginning to be tapped, except in a few remarkable classrooms.

A practical example: Add 68 + 27 <u>in</u> <u>your</u> <u>mind</u> right now, noticing your thought patterns.

How did you get 95, the answer? Did you picture the two numbers on a kind of blackboard of the mind, add the 7 and 8, carry 1, then add the 6 and 2? Even though you were visual after a fashion (visual-symbolic), that was a fairly left-brained approach. You were reproducing the pencil and paper rules (algorithm) in your head.

Or you may have rounded 68 up 2 to 70 then added 70 + 27 = 97 then adjusted the answer downward by 2 to 95. Or you may have added the 60 and 20 part, then noted that, because of anticipated carrying the answer would be in the 90s and end in 5, because 7 + 8 is 15.

Others add 68 + 20 = 88 then add on the 7 by first adding 2 to get 90, then the other 5 to get 95. All of these routines are more right-brained because they utilize non-rule-bound relationships between parts of numbers, and because they feel or visualize patterns.

A more dramatic example: Once a friend of mine was involved in a math bee in elementary school. He and a girl were the finalists, set to play off at each end of the blackboard. The teacher dictated the calculation problem, and both took off. My friend soon perceived that his chalk was moving about half the speed of his opponent's, and his left brain rationally concluded that he would surely lose. So he was inspired to just *stop writing*. He stood back, looked at the problem, then wrote the answer on the board. He won!

What happened here? The left brain has been trained in school to take over most intellectual processes. Whatever relationships, subtle wholes, and interactions the right brain perceives are drowned out by the intense chugging of the systematic left brain. Shortcuts, analogies, and connections are all overlooked as we slog through it all the long way. My friend's left brain simply surrendered to the inevitable defeat it was heading for (some left brains are less willing to give over the floor), and this left a wide-open road for his right-brain input. Presto! The answer was already developed and ready.

Science has long been mystified by the feats of autistic savants, who can barely function in most of life's roles, but who can do almost instant mental calculations that would take a calculator a fair time to perform. Undoubtedly, rather than just using left-brained digit-crunching, the savant is perceiving subtle numerical relationships during these feats.

Now, as a footnote, I might seem like I'm giving the left brain a bad rap. Perhaps I'm overcompensating for the tremendous

The secret is to bring everyday-life thinking and cleverness into math problem-solving.

emphasis the left brain receives in our society. Medicine fixates on curing symptoms (details) rather than looking at causes in the whole body, mind, and emotions. Political solutions react to crises (details) rather than doing long-range, broad adjustments to society. Businesses focus on the one-dimensional, linear bottom line (details) rather than the long-term effects of their activities on society and the environment. And education focuses on learned facts (details) rather than the interrelated skills and competencies of solving problems and doing complex projects.

One simple way to tap the powers of the right brain in your next lesson is to give an *overview* of the lesson first rather than plunging into detail. This single technique will orient your students, create mindset, and greatly relieve those who are more right brained.

Count among your right-brained students those many (usually boys) who are diagnosed with ADD or ADHD. Jeffry Freed makes the point in his book *Right-Brained Children in a Left-Brained World* that the label is over prescribed and that the small percent of youths that are truly ADD are universally right brained and hypersensitive with low impulse control. Rather than thinking of this as a deficit or curse, he considers it just a difference and a gift that can be tapped for creativity, intuition, and spatial problem solving. Such students have very acute hearing, smell, and sense of touch as well as uncanny intuition that is heavily stifled by left-brained demands around them. Consequently, any of these may severely distract these students: buzzing lights, outside noises, classroom cleaning odors, being touched. Meanwhile, they struggle to get sequential steps down on paper while their intuition soars unspoken. Freed has "freed" many of these students by helping them tap into their mental computation abilities and numerical intuition while quieting environmental stimuli and toning down their "sequential obligations." His book is a valuable read for teachers and parents of these students. ADD and ADHD diagnoses often mask dyslexia. Multi-symptomed, dyslexia garbles symbol processing (hence reading and math). Frustration and low esteem result. Dyslexics' strong visual creativity undermines steady symbol perception. MI instruction definitely helps, but it's not a cure. For insight and an amazingly effective method of curing using simple exercises, see Ron Davis' dyslexia website and book (Resource Bibliography: Davis).

To summarize this chapter, the intent of math education should be to enable students to do mathematics using *both sides* of their brains. The predominance of left-brain thinking in our schools and textbooks subtly convinces teachers and students that math is only for the left-brained. In fact, many students and adults automatically lobotomize their right brains whenever they come into contact with numbers and math thinking suffers.

Learning Styles and Math

TEMPERAMENT AFFECTS NUMERACY

For decades teachers have been aware of the three "learning styles" that relate to the three modes (auditory, visual, kinesthetic/tactile) by which a learner takes in information. Each learner has a dominant style. A person who can learn from sentences on a blackboard would be called visual-symbolic, while someone who could learn from a diagram would be visual-spatial. A child who can grasp a number fact when it is traced with a finger on her back is tactile. The child who can learn by <u>doing</u> the tracing is kinesthetic. Someone who can learn from a taped lecture is auditory.

This classification system of input channels is still handy and has even been given more refinement by Dawna Markova, Ph.D. *(See the Resource Bibliography: Markova.)* However, "learning styles" has come to mean much more. While those listed above are modes of input, we can speak of the ways in which students engage and process information in their learning environment (or the world), regardless of which *channel* they used to input it.

I find it simple and helpful to consult a system developed by Anthony Gregorc, an educational researcher. *(See the Resource Bibliography: Gregorc.)* It contains four different temperament/processing styles. The combinations possible between a major and sub-major style lead to about ten variations of learners. The styles will be presented not in Gregorc's more left-brained categories and charts, but in a more right-brained manner — using the symbology of four animals, and adapted to math learning.

There are four major animals that seem to characterize each of us — student and teacher — as we operate in life and in math. Generally, we can each be represented by one, or a mix of two, of these animals. Each animal represents a combination of temperament, operating procedures, organizational preferences, and abilities that all seem to come in a kind of unified "profile."

The Beaver

The first animal is the **beaver**. I'll give the animal some human traits as I describe it. The beaver is a very practical animal that gets things done. It's built to produce a product – a dam. It's got teeth to cut trees, powerful legs to drag them, a tail to pat mud, waterproof skin, and webbed feet to swim. It goes "by the book" (i.e. it follows the plan for dams). If there were a book on dams, it would buy it and follow it to the letter!

Beaver teachers like to keep things in order, to plan, to line students up in straight rows, and keep precise records of their performance. They're very reliable. They demand neat papers all in the same format, and their daily classes are predictable. They don't like surprises. More enlightened beavers like "hands-on" reality, input from the senses, concrete information, and practical experiences, but less inspired beavers often settle for worksheets and percentage grades as the only concrete reality in the school setting. At their best, beavers are reliable organizers and a rock to lean on. At their worst, they can be reactionary, oblivious to the deeper needs of people, and into a rut of meaningless routine.

Beaver students like regularity, accomplishment, products, a gauge of the payoff from each activity, and a framework for keeping track of all imparted information. They love practical, hands-on experiences that are well-defined and organized, so they need to be strongly encouraged to explore and experiment and should be regularly exposed to very concrete math applications.

Incidentally, surveys performed about twenty years ago indicated that about two thirds of teachers in first through twelfth grades were high in beaver qualities! I doubt this has changed significantly in the 2000s. This means textbooks, curricula, and classes will tend to be this way, too. The chances, then, of students having been taught math in the past by several beavers are very high. Whether students reacted negatively to them or loved them, their ways of learning may be very conditioned by this style.

Beaver students can settle for having lot of papers as their concrete products; they would shine more in math if asked for real-life physical projects.

The Owl

The second animal is the **owl**. We've all read of the "wise old owl" in children's stories. He generally wears a pair of glasses. An owl is a night animal, not distracted by the emotions and activity of daytime. Just look at those unblinking black eyes! He's an analytic thinker, organizing ideas into an orderly matrix — kind of how the beaver organizes logs into a dam.

The owl's clear mind cuts through the maze of detail to penetrate to the essential thought. She's "in her head" most of the time and finds emotions to be something to "think about." Sometimes the owl teacher's ideas don't work out in practical reality as well as they do on the drawing board. She may be overly dependent on words, creating too many distinctions to make a quick decision, and may resort to sarcasm, when angry. At her best, an owl can clarify the worst informational muddle. At her worst, an owl may consider an idea or "-ism" more important than people, and talk over everybody's heads. A typical student comment on such an owl is that she is very well versed in her subject, but she can't get it across to "ordinary students."

Owl students need to know "why" something works and not just how to do it. They like their minds tickled, regardless of whether the

Owls are often not motivated by the practicality of mathematics. They are more turned on by amazing number behaviors or the analysis of why things work.

information is "useful." They need time to reflect. They need to be challenged to get to the bottom of an idea. They enjoy knowing the fundamental organization behind the concepts being presented, not just that they happen to work to get a correct answer.

The Dolphin

The third animal is the **dolphin**. Dolphins are playful, artistic, imaginative, subtle feelers. They exist in a water (symbol of feeling) medium. Animal dolphins have been known to rescue humans and other animals, so I call them "humanitarians." Human dolphins "care" about what others forget to care about — like anniversaries, thank you notes, and the like.

Most people would call dolphins "colorful" animals, in a metaphorical sense. Human dolphins like color, tone, and artistic expression. Little thoughtful things are important to them. They *feel* into other people's insides. Their motto is "trust me" (e.g.: "I know that the person who left this note saying she can come for an interview is the person we will hire. How do I know? Trust me, I know.") Their knowledge comes from an unknown inner source. A dolphin spouse might council a beaver mate, "Don't trust that person on this business deal." The beaver asks, "Why not? The papers are in order, the explanations make sense, the investment looks foolproof." The spouse says, "The man is dishonest. Trust me." This intuition usually proves correct.

Dolphins living in the concreteness of our goal-oriented, time-organized, money-driven society sometimes think they are aliens. They have the opposite of those values in their heart of hearts, living in the present and placing highest value on feelings that no pricetag describes. At their best, they create warmth and color and honesty around them. At their worst, they are self-absorbed, undirected in their energies, and disorganized.

Dolphin teachers are rarities at universities (owl havens); they cluster in the humanities in high school and junior high, and are sprinkled around the elementary grades. They congregate at the primary levels, kindergarten to third grade, where they can treat their classrooms like a family, use bright decorations, and enjoy a context of fantasy in the class.

Dolphin students need a warm relationship with the teacher in order to learn, and they enjoy personal interactions with other students while learning. They want to know the personal side of the ideas, the juicy history, and the present personalities involved with them. Being so intuitive, they enjoy mysterious and colorful ideas most, so geometric design, number patterns, and mathematical mysteries can be their inroad to math. (*See the Resource Bibliography: Wahl.*)

Dolphins are usually the least reached by math classes. They don't relate well to cut and dried, black and white, lifeless symbols. They need flesh and blood, color, and interaction.

The Fox

The last of these representational animals is the **fox**. Foxes are clever, devising different ways to get into the chicken coop or to elude the hunter. They like adventure, so human foxes like new methods to solve problems. They enjoy risk, and by extension we can say they enjoy pioneering "state of the art" things or ideas. They like innovations and trends. They shoot from the hip rather than through extensive deliberation, and they can deal with several situations at once. They grow impatient with a slow, steady, methodical approach to learning, thus they often get on the wrong side of a beaver teacher.

To foxes, regulations are made to be stretched to fit new situations. They are claustrophobic amidst too much structure and too many rules. They are entrepreneurs, taking intuitive readings from unfolding situations to reap an advantage. To them, dolphins are too thin-skinned or airy-fairy, owls are tediously theoretical, and beavers are uncreative clones. At their best, foxes are stimulating and forward-thinking; they come up with creative solutions. At their worst, foxes are dilettantes and quitters, going with trends but avoiding depth.

Fox teachers are a small minority (about 5%) in teaching. Perhaps they are repelled by the general lack of creativity that institutional teacher training has fostered. Or maybe they run afoul of all the rules, consistency, and necessity for documentation that standard teaching methods require.

The fox student thrives on challenge, variety, surprise, and risk. He can accomplish more than one goal or project at once. He needs to be encouraged to follow through on tasks before starting new ones, but his spontaneity and love of new approaches must be fed. Daily lessons should contain some element of surprise. He should be given open-ended tasks, unusual challenges, and a choice among options as part of his homework fare.

Foxes will create surprise, risk and variety in your classroom — negatively if you never inject it into your lessons positively. To them, endless repeated routines are like sitting still while fingernails are dragged across a chalkboard.

Best use of this learning style information

Knowing that your students have wholly different attitudes and approaches to your curriculum will begin to have some effect on your lesson plans. However there is more you can do. First, you should reflect on and determine your own animal(s) so you know your bias when you teach (because, mysteriously, students with *our* learning style always seem smarter!). Then it would be helpful for each student (and you) to know what that student's style is, as a way for both of you to understand and alter the process of learning, especially when it is clear that a student is out of sync with your approach.

Regardless of age, students can learn about their styles through the animals. Usually, the older a student is, the more deliberate will his choice of animals be. Be aware that younger students will at first tend to simply choose animals they have always liked rather than the ones that represent their style!

I find it helpful to describe each animal in detail, then ask which students identify with which animals. Then I further interrogate them about *why* they chose the animals. This helps to clarify the styles and refine the choices. Be sure to leave open the possibility that some students may feel an affinity with **two** or **more** of the animals' styles.

If a child relates to the characteristics of two animals, even after refinement of the meanings, it is helpful to try to determine which animal, if any, is the most dominant. Occasionally a child will possess versatility in his relationships, being friends with different "animals" and will operate with characteristics of all at different times. She will feel like *all* the animals ring somewhat true even after refinement of her choice. This is usually evidence of an adaptability that can be helpful in life and learning, the only liability being less of a focus of interests or vocation.

You may find it interesting to have your class arrange by homogeneous animal groups, assign a task, and help them observe afterwards how differently each group went about it. You can also emphasize to them how each animal might take a little different approach to class work, homework, extra activities, and grades. Let everybody pool their observations in a class discussion. An example is that the beavers might get involved with production and accuracy, the owls with clear reasoning and concepts, the dolphins with tracking how everybody is doing and getting support when help is needed, and the foxes with challenges and looking for shortcuts. Hopefully they will begin to see how each animal could be an asset in their cooperative groups. *(See Chapter 6, "Cooperative Learning", for how to set up this mode in class.)*

Depending on your style, teaching some of the animal types will be easy and others will be more challenging. Your job is to have at least something in the style of each animal in each lesson. This takes time, but try, a bit at a time, to bring in approaches for those animals that seem most *opposite* to your style.

Students of each style need to broaden to other learning approaches, but deserve to learn in their own style a lot of the time.

Cooperative Learning

THE WHOLE SUMS TO MORE THAN ITS PARTS

Teaching challenges abound in today's schools. Our society and schools are plagued by a sizable percentage of youth demonstrating uncaring attitudes, violent behavior, suicidal alienation, prejudice, and desire for belonging *of any kind*. An argument can be made that the competitive and individualistic nature of societal and school structures help breed these problems. In addition, as of 1997 national education laws mandate inclusion of more children with learning differences in general education classrooms after generating Individualized Education Plans (IEPs) for each student. Furthermore, more minorities with unique needs appear in the classrooms.

Cooperative learning is a powerful tool now being used successfully in many schools to help deal with all of the challenges mentioned above. For example, it has been found that minorities respond especially strongly to this model. David and Roger Johnson researched, refined, and popularized it, and further structures have been developed by Spencer Kagan and others. *(See the Resource Bibliography: Johnson and Kagan)*

Cooperative learning is not a quickie cookbook ingredient that a teacher can install overnight. Rather it is a style that can be grown in a classroom over the years, and it is fed by cooperative activities among faculty and administration. My short description here is not a manual for implementation, but rather an overview whose intention is to pique interest and encourage you to *try* cooperative learning in your mathematics teaching.

When they hear the term, teachers often think they have used the cooperative learning model many times, but few have. Teacher manuals for the latest math books use terms like "cooperative groups" but rarely include all of the elements that will make the cooperative exercise truly effective.

"Cooperation is *not* having students sit side-by-side at the same table to talk with each other as they do their individual assignments. Cooperation is *not* assigning a group report that one student does and the others put their names on. Cooperation is *not* having students do a task individually with instructions that those who finish first are to help the slower students. Cooperation is much more than being physically near other students, discussing material with other students, helping other students, or sharing material among students, although each of these is important in cooperative learning" — David & Roger Johnson.

A full cooperative learning experience requires:

1. Clearly perceived positive interdependence

Negative interdependence is easy to find in schools — competition and rivalry. Its opposite can be structured by a teacher in several ways. Two or more of these ways incorporated in an activity add significant breadth and strength to the learning experience:

➤ Positive goal interdependence. "All members of the group must be able to outline the solution procedure for the problem."

➤ Positive reward/celebration interdependence. "When all members can write out their version of the solution correctly, all will receive bonus points and extra computer time."

➤ Positive resource interdependence. "Each participant in the group has been given data relating to the problem."

➤ Positive task interdependence. "Each person's data must be analyzed and put in a form the group can use for solving the problem."

➤ Positive role interdependence. "Make sure you have a recorder, a summarizer, and a checker of understanding in each group."

Most of us adults had little schooling in how to cooperate. Don't be discouraged if your first attempts to foster cooperation in children create a certain amount of disorder.

2. Face-to-face Promotive Interaction

Students need to be seated in a way that promotes face-to-face exchange. These exchanges have multidimensional benefits like assisting, exchanging resources, giving feedback on ideas, challenging assumptions, encouraging one another, and showing trust. Cooperative learning is not just about academic learning but about social learning as well.

3. Individual accountability/ personal responsibility

There are ways to encourage everyone in the group to contribute fairly. Here are two:

➤ Keep the size of the groups small (two or three usually, sometimes four).

➤ Each student must be assessed individually (written, oral, by teacher observation, or by student checker) with regard to the joint project.

The main principle is that students learn (and sometimes perform) together, then perform alone.

4. Interpersonal and small group skills

A component often overlooked in so-called cooperative activities is instruction in group skills. Watch a committee of adults struggle to solve a problem, and it is likely you will see deficiencies in group and interpersonal skills rear their ugly heads and obstruct efficient progress. Most of those adults have probably not had any instruction in cooperative work.

Similarly, your students will not automatically become effective cooperators. However, with instruction, they will multiply their accomplishments within groups. Some overall skills outlined by the Johnsons are: "(1) get to know and trust each other, (2) communicate accurately and unambiguously, (3) accept and support each other, (4) resolve conflicts constructively."

Each of these has subskills that help bring them about. Four examples of subskills, each at increasing levels of sophistication, are:

➤ encourage everyone to participate,

➤ paraphrase another member's contribution,

➤ summarize out loud the group's results thus far, and

➤ integrate several viewpoints into a single conclusion.

How are these subskills instructed? I'll give you an example. The Johnsons recommend constructing a T-chart (with student input) for each skill you're emphasizing during that day's task. On its left is the body-language that accompanies the skill. On the right of the chart are verbal cues that indicate the skill going on. Explain and discuss its content with the class.

Here's a group-skill example:

Encourage Others' Contributions

Looks Like	Sounds Like
Friendly eye contact with quiet person Listening when a quiet person contributes	"Did you have something to say?" "Who else has an idea?" "Let's hear something from each one here."

Then, after instruction in the skill, you ask that each participant strive to practice it twice that day. You monitor them in this regard and record when you see it. They're encouraged by this attention to try it. Then check back at the end of the day to see if all were successful in using it.

5. Group processing

This fifth ingredient is very important. Group members are asked, after achieving their goal (like solving a difficult problem), to reflect back on how they got there, what the key turning points in cracking it were, what each contributed to the solution process, how many times each used the day's emphasized cooperative skill(s), and so on. These questions should be asked one at a time with a few minutes given to the groups for each reflection. The groups can then contribute their insights on their processes and the problem solutions to a fertile whole-class discussion.

The Johnsons' meta-studies of the many experiments done with cooperative learning in the last 90 years reveal that this model is at least as effective as individualistic and competitive models in all settings and with all subjects, and it is usually far more so. Additional benefits include attainment of higher-level reasoning and idea processing, more new ideas and solutions, greater transfer of what is learned to other situations, greater retention, far more interpersonal skill growth, greater self-esteem, and higher regard for others.

Many of the activities in this book are ideally suited for use in a cooperative setting. Give them a try and check the Resource Bibliography for further reading on this valuable technique. Don't expect to master it overnight, but congratulate yourself on each small step you make to increase cooperation in your classroom and in your faculty.

Spencer Kagan has concretized patterns for student interaction as structures. Together, teachers and students develop a repertoire of several of these structures. So when the teacher asks the class to use the "Think-Pair-Share" structure to solve a problem, students know they will work independently to write down their initial approaches to the problem, then find a partner, share ideas with that partner, then attempt to complete the solution together.

If a teacher announces the "Find-Someone-Who" structure for skill or terminology practice, each student receives a worksheet with the instruction to "Find someone who knows the answer to..." for several questions. Keeping a hand raised each seeks an unpaired partner (with raised hand). Only one answer is sought from each student encountered. That person writes it, along with his name, on the student's worksheet if the student accepts the answer. When students finish they become helpers by sitting down and becoming a "found person" for others who can ask them any question. This is a great way for students who originally know few of the answers to become a resource for others, now that they too are "someone who knows."

There is no perfect cooperative learning teacher — only one who is regularly improving in the use of the model.

Dealing with Math Anxiety

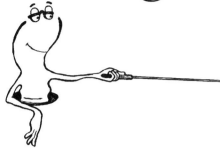

FEELING GOOD MEANS LEARNING WELL

It is no surprise that math anxiety and avoidance are widespread among U.S. students and adults. Forty countries' math education programs were scrutinized as part of the Third International Mathematics and Science Study (TIMSS), whose moutainous data was analyzed in 1996-97. Additionally, 500,000 students ages 9, 13, and 18 were tested. Results revealed that in a world comparison U.S. math training has fragmented goals and limited conceptual depth. U.S. 4th graders' math performance is just above the world average, whereas the 8th grade is mediocre and the 12th grade is abysmal.

These students move on to join the adult world with perhaps 75% of its members experiencing math discomfort that ranges from simple avoidance to actual phobic reactions like sweating palms and blanking mind. This syndrome often starts in grade school and usually only gets worse, unless an understanding teacher and good learning experiences intervene.

This means that your classroom or child will probably have a fair share of math anxiety that needs to be addressed. (After primary grades, cases of math anxiety will generally be even more numerous each grade from four through high school.) To simply bring out the balloons and party hats and say, "Math is fun!" will not relieve the deep-seated pain and defeat some students experience with numbers. Successful work in this area requires a multipronged approach.

First of all, teaching through the multiple intelligences is an automatic pain-reliever, because now a student whose mental receptors, for example, are tuned to Channel S, the spatial channel, will not just receive static when a teacher is broadcasting only on Channel L, the linguistic-verbal channel.

Whatever a student's strongest channels, it's also important to twist the knob to Channel I, the intrapersonal channel, and help each student acknowledge his or her past fear and pain around the subject of math. This can go well beyond just a general gripe session about past math experiences, though even that can be a start. A student will unload a lot with a little sympathetic listening (e.g., feeling stupid on tests, being embarrassed at the blackboard, being told he was slow in math, feeling afraid to ask questions, etc.). This in turn can give you many insights into how to proceed with your students' math education so as not to "pour more salt on their wounds."

But there's more that can be done. A student needs help to reflect on where this math fear came from in the first place, and that it does *not* mean her brain circuitry is faulty. Here are some of the several seeds that could have grown into a case of math anxiety and subsequent failures:

➤ The student may have been pushed into math symbols too fast in the primary grades, without enough playful, exploratory experiences. Feeling "out of it" early left gaps snowballing through the grades, and she has never caught up.

➤ He may have subtly picked up a case of number fear from a parent, who had his or her own wounds and issues with math — then this family belief that math is a trial became a self-fulfilling prophesy in the student's worldview and class performance.

➤ A difficult classroom situation or insensitive teacher in the past may have "turned her off" or "clammed her up."

➤ The student may have changed schools several times and missed certain vital math jargon that the others in the class knew, so he fell behind and called himself stupid in math.

➤ The student may have a different learning style or intelligence profile than that used in most of his math classes, so the material seemed incomprehensible, and he turned off.

The main point here is that the student must see a bigger picture about how these math fears are caused and that they are not a sign of low intelligence. She must come to trust that math achievement and enjoyment is possible for her. She must know you believe in her and that you assume her math fears were just a healthy response to adverse influences. You firmly believe that with good math experiences she can blossom into a reasonably competent math student.

There are many layers and multiple causes of math anxiety in an individual. We can at least attend to the obvious ones.

Lowering math anxiety in the classroom

The first step in reducing and removing student stress is to acknowledge that it exists in some of your students already, and to discuss it openly with the students. If you have had some math anxiety yourself you might wish to let them know this and share with them how you have dealt with it. You could indicate your desire not to increase it for anyone in the class, and you ask for their help in letting you know, even with a signal, like making an "A" with their fingers, when they are slipping into the black hole of fear, anger, confusion, frustration, mind-blanking, or self-put-down.

Next, you need to compassionately and honestly monitor yourself for ways that you may engender anxiety as you teach math. Here's a little check-list of anxiety-producers to start you off. You can add to it as you observe your individual style.

- ➤ Putting a student on the spot to answer or explain as the class waits.

- ➤ Sending a student to the board to show work (unless he expresses a desire to do so).

- ➤ Overly praising or rewarding right answers and expressing mild disappointment (or even neutrality) with wrong ones.

- ➤ Pronouncing answers "wrong" without noting the many right steps in them.

- ➤ Placing emphasis on fast work and limited time in class or on tests.

- ➤ Cutting off a wrong oral answer with no attempt to track how the answer was obtained, or no acknowledgement of what part of the process was right.

- ➤ Discouraging certain seemingly naive questions or treating certain questions as irrelevant or irritating.

- ➤ Publicly or privately shaming a student for inadequate work.

- ➤ Loudly explaining a task to a student who is behind on the work.

- ➤ Teaching mainly to the fast students (or to the boys or to the girls).

- ➤ Expressing impatience or exasperation with student slowness in grasping a concept.

- ➤ Giving complicated or convoluted explanations for a process when it is clear that only a few are grasping it. [Instead, acknowledge the bad start and go back to build up understanding from a simpler stage.]

- ➤ Seating students according to their abilities.

- ➤ Posting grades on tests or making students publicly chart their progress.

- ➤ Assigning tasks that are overly daunting, complex, or time-consuming.

- ➤ Assigning math problems as punishment or detention.

- ➤ Publicly or privately referring to a student as poor or slow.

If you see a student sliding into the black hole during a class exercise, try to speak privately with her and bring out what feelings she is experiencing. If most of the students seem "out of it" during a task, stop the task and have a short discussion of the "feeling tone" in the room. Initiating this can sometimes feel threatening to you as teacher because it seems to invite the comment, "I hate this!" It is important to get past this to a classroom climate where a student feels free to express a feeling.

Your goal is to create a trusting classroom atmosphere where students help and look after each other, where they are candid about what is going on with them, and where they see you as a warm, non-threatening advocate for their learning. Then, as you use the more varied and fun teaching tools discussed in these pages, and you excite your students, math anxiety will fade like a bad dream and be replaced by smiles and confidence.

A further note: A history of learning disabilities, vision problems, difficulties in the home, poor nutrition, food allergies, missed early math information, or trauma from previous school experiences can all be contributors to a student's severe math anxiety problem. These can't be solved overnight but they should be checked for and their remedies undertaken by appropriate personnel. However, I have seen students with very negative attitudes about math and having some of the problems I've listed, gain bright happy faces and start really clicking when they are learning ideas on their wavelength. This is the possibility that teaching through the Multiple Intelligences holds for you!

Severe math anxiety leads to diagnoses like "discalculea" ("can't calculate"), ADD, ADHD, dyslexia ("can't read"), etc. These pretend to "explain" bad math performance but suggested remedies are often weak. Such labels can become harmful catch-alls, masking mediocre teaching. I have personally seen multiple intelligence math teaching create dramatic turnarounds with numerous "diagnosed" students.

While some conditions are true "learning challenges" like dyslexia, ADD and ADHD, (see the discussion and resources on the bottom of page 36), many are not neurological in origin. For them, an early learning difference or slower development curve can by amplified into a disability by impatient parents or one-size-fits-all school expectations. Understanding this dynamic and  applying a good dose of MI teaching can reverse many symptoms. Assume non-neurological first but read the references on p. 36 to spot symptoms.

A final note: On the next page there is another way to get classroom or individual discussion of math anxiety going. Use the inventory there to have students rate their degree of math anxiety. Then go over the results afterwards, either in class or individually, and see if they agree with how it has scored them. If you have younger students, you may want to discuss each item first then have them put down a number.

How Do I Feel About Math?

Put a number from 1 to 5 next to each of these statements according to whether it is...

Almost never true, or you have little feeling about it = 1
Sometimes true, or you have some feeling about it = 2
Usually true, or you have a definite feeling about it = 3
Almost always true, or you have a strong feeling about it = 4
Always true, or you get a strong emotional reaction = 5

1. I feel an urge to play around, socialize, or stare out the window when math starts. ☐

2. When I meet students who love math or do it well, I either think they are a little weird or I envy them. ☐

3. If I am sitting with two students who are talking about math I have an urge to get out of there or do something else. ☐

4. When math starts I get a physical reaction in my body, like tightening, or tiring. ☐

5. Being asked to "go to the board" to explain a math idea in a class — even for math I am able to do at my desk — scares me. ☐

6. I'm not sure I can trust my answers, even on simple problems. ☐

7. I have a hard time sitting down to start math work outside school. ☐

8. Math never seems to stick, and after I learn it or even get a good grade on it, I still don't think I know it. ☐

9. When I'm around a hard math lesson or task, I feel

angry ☐   scared ☐   stupid ☐   tired ☐

helpless ☐   blank or fuzzy-brained ☐

10. In my body, challenging math problems give me:

upset stomach ☐   headache ☐

sweaty palms ☐   drowsiness ☐

Now add up all of the numbers you have written: _____

If you wrote no **5s and no more than one 4,** and you got **24 or less**, you probably feel fairly OK about math, but might need more practice or a little different instruction.

Otherwise here are your math anxiety estimates:

25-35 Some math discomfort and anxiety.

36-45 Quite a bit of fear and discomfort with math.

46-55 Very anxious about math. Talking about and working on this with your teacher, and maybe with another adult you trust will help you a lot.

56-85 You are just about paralyzed by math! You have a lot you can gain from talking it over with your teacher and another adult you trust. It would also help to have your instruction or testing methods changed to make it easier for you to feel comfortable learning math.

Assessment Ideas

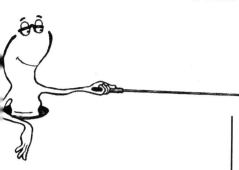

ACTIVITIES CAN MEASURE MANY PROCESSES

I once saw a cartoon of a managerial person sitting behind a desk speaking to a job applicant seated on the opposite side. The manager is saying, "After examining all the information we have about you, we have concluded you are good at one thing — taking tests." Standardized testing is a multi-tentacled giant that has gained even more momentum from federal "No Child Left Behind" legislation and state politics. Countering these is a strong parent reaction and widespread teacher innovation, resulting in more performance assessment, appropriate rubrics, self-evaluation, and student portfolios.

Assessment has often been used as a carrot, and just as often as a stick. It needs to evolve into a simple, less "loaded" feedback mechanism for students and it needs to become a flexible method of informing parents of a student's in-progress accomplishment toward the goals he plans to reach for the year. The philosophy and Activities in this book are generally self-motivating, virtually eliminating the seeming need for carrot or stick, so let's look more closely at how you can use assessment for its feedback and information functions.

For either of the two assessment functions we can distinguish *formative* evaluation from *summative* evaluation. Formative evaluation gives students feedback while they're still producing and provides the opportunity to upgrade their performance based on this feedback. Summative evaluation, often overused by teachers, grades the effort afterwards, when nothing can be done to change it.

I have found it wise to institute formative *reflection* among my students. This allows them to evaluate the caliber their work has reached while they're still immersed in an activity. By discussing your (and their) criteria, both before and during the activity, you can then solicit their assessment of their own partially completed efforts by asking them to look for ways to improve their work and by providing examples of excellent work that your past students have generated. Your students can then give themselves in-progress *scores* using "rubrics" extracted from the agreed-upon criteria, which can contribute to marks reflecting the degree of mastery of the learning. (Some examples of rubrics will be presented below.)

Experts like Glasser believe that much of the assessment function can ultimately be turned over to the students. *(See the Resource Bibliography: Glasser.)* They must first learn to spot "qual-

ity," then how to evaluate their work and, from this evaluation, how to make an effort to improve their work. If this process is done correctly, it will encourage your students to strive for "quality" work, work that is well above — in fact in an entirely different category than — just "good" work. Quality work creates satisfaction, absorption, and meaningfulness without imposing these from outside in the form of a grade, reward, or teacher's approval.

You can provide summative evaluations to parents (if those are required by your school) by incorporating into your usual grade reports the final assessments of student math activities that include rubric *scores* (not letter grades) in various process and skill areas. (Examples will be given below). If grading procedures in your school are tightly or traditionally defined, rubric scores on the activities can be converted to an A, B, C, D scale and factored into your regular grading system. But be aware that traditional grading systems need redefinition; help your school move toward that. It's a fact that earning traditional grades on the activities can take away some of the sense of free exploration and excitement about learning that multiple-intelligence activities generate.

Ideally, summative assessment for the benefit of parents and administrators would be done with multi-intelligence presentations and projects coordinated with a portfolio system (see Resource Bibliography: Stenmark). This involves assembling in an individual portfolio the results of larger, integrative projects, feedback on student presentations of these projects, and good examples of a student's in-class work. Included with that work can be the student's, and your, rubric ratings and comments. Some portfolio pieces should even include scratch work and preliminary explorations to show the progression or development of final presentations.

Remember that discussion of the criteria with students is essential to keep rubrics from becoming a "top-down" system like most assessments.

Now for a discussion of rubrics.

Because the Activities in this book, and your multiple-intelligence teaching, use *process* skills that include problem-solving, data processing, exploration, and inquiry, I recommend creating an uncomplicated set of rubrics for assessing these in a formative (or even summative) way. You may want to refine the set of rubrics that I am offering below to meet your exact needs. A class discussion of your suggested rubrics, in which changes and additions are proposed by students, should be held in advance of doing any activities. Then students will be familiar with the standards and better able to apply them in their self-evaluations. You can request that students, upon finishing an activity, self-rate their work as individuals, or as a group, in the categories listed below. As trust builds during the year, members of a cooperative group may give input to each member as he or she does a self-rating aloud. I have seen this work well but only if it is completely voluntary.

Here are some suggested learning processes and their ratings (applicable to a student, but adaptable to a group). You may wish to emphasize only one or two processes for any particular math activity. The ratings go from a high of 3 down to a low of 0 (note that $\frac{1}{2}$-step increments, e.g., $2\frac{1}{2}$, are possible).

A. Understanding the task

3. The student shows curiosity, perseverance, and self-motivation in attempting to grasp the task and, if stuck, formulates good clarifying questions to ask the teacher and other students. The student can describe in his or her own language all the important elements of the task and can prescribe what mathematical processes need to be done.

2. The student shows some curiosity and grasps the requirements of the task after some wrong turns. He or she seeks help with specific questions about the activity and shows some perseverance in carrying out the task. The student can describe, though not always precisely, what is being required and what processes are needed.

1. The student is somewhat baffled by the activity and can't see the big picture of what it is about. He or she needs guidance and encouragement to figure out the activity and its requirements. He or she makes some progress but finds it hard to coherently convey the point or processes of the activity.

0. The student essentially cannot respond to the directions, produce results, coherently discuss the activity, or continue the task without constant guidance.

B. Completing the task thoughtfully and thoroughly

3. The results are shown in a coherent, detailed, and complete form with clear indications of which part of the activity they relate to. Comments or labels help organize the work. Some suggested extensions have been tried and reported.

2. Results are organized and reasonably complete, but may lack comments and labeling. No extension, or just part of one, has been tried and partially documented.

1. Results are partially complete and in a disorganized form.

0. There is little sign of consistent work or meaningful results.

C. Quality of collaboration (if cooperative groups are involved; see Chapter 6 "Cooperative Learning")

3. The student utilizes several of the emphasized collaborative skills. The activity progresses with the benefit of exchange and teamwork. The student can relate some results to having worked with others in the group.

2. Utilizes one or two of the emphasized collaborative skills. Participates in collaborative activity for a significant part of the time. Some significant results or documentation occur as a result of collaborating on the task.

1. Occasional collaboration occurs, but it primarily results in using someone else's results. Some student activity detracts from cooperation in the group. None of the emphasized collaborative skills are practiced.

0. No solid or lasting signs of collaboration during the task. Behaviors detract from a cooperative atmosphere.

D. Accuracy of answers (No need to turn the Activity work into "percent right".)

3. Almost all answers are accurate and complete.

2. A strong majority of answers are accurate and complete.

1. Around half of the answers are accurate and complete, while other sections are omitted or done incorrectly.

0. None or few answers are correct.

E. Understanding and application of the mathematical concepts of ... [list specifics].

3. The work shows clear ability to explain, select, and apply these concepts.

2. Most of the work with these concepts is done with some understanding, but some is done with mechanical imitation that the student is unable to justify. The student knows which operation or procedure to use a majority of the time. The student needs some prompting to be able to select some of the procedures used.

1. The student is operating mechanically, often without the ability to select or justify the procedures.

0. The student has few clues about which procedure to use or, when a procedure was suggested, cannot carry it through.

There are other processes or skills that you may wish to assess with respect to the Activities at various times. For example: problem-solving ability, use of multiple-intelligence problem-solving strategies, and research skills. *(See the Resource Bibliography: J. Westley as well as Activity set 14 on pp. 14-1 to 14-19 called "Get Really Smart: Crack Problems With All Your Intelligences.")*

A main idea to remember is that seeking the *intrinsic* rewards, like the joy of learning math, is to be emphasized over the *extrinsic* reward of a good mark. Another assessment *must* is to help students internalize the criteria of quality work, then to encourage them with examples and involve them in discussions of criteria to produce quality on a regular basis. *(See p. 24 for a short discussion of "metacognitive training," a successful approach for helping students to get in charge of their learning goals.)* Bear in mind also that students will look to your teaching as a model of the very qualities you assess: curiosity, collaboration, clarity, preparation, creativity, completeness, and involvement in the learning process.

✚ ━ ✕ ÷ Supercharged

POWER PACKING FOUR OPERATIONS WITH MI

Long ago I heard a true story from an educational researcher who asked a schoolgirl how she knew when to add, subtract, multiply, and divide. She replied something like, "When there are more than two small numbers about the same size, I add; when there are two larger numbers roughly the same size I subtract; when I'm looking for a big answer, I multiply; and when one number is small and the other large, I divide." This kind of guessing and pseudo-rule-making behavior often surrounds the four basic operations, making the solving of story problems seem like a mix of shooting dice and magic.

It's not uncommon for students to learn the tables for all of the operations and still have a fuzzy notion of what they're used for. A good test of whether a student understands the operations is to ask this kind of question: "Give a real-life problem (involving cars, gardens, horses, money, CDs etc.) in which a person would do this calculation: 385 - 137 = 248."

Many (though not all) students will be successful on this subtraction example by specifying a *take-away* situation. Only the more insightful students, however, would create a problem of the *comparison* type: "Calvin collects aluminum to recycle. He has raised $385 to buy a trail bicycle. His friend Jean is doing the same thing but has only raised $137. How much more has Calvin raised?"

But while only some will falter here, many more stumble on similar questions involving multiplication and division of whole numbers, and they scatter like bowling pins before questions involving fractions or decimals: "Give a real-life problem in which a person would do this calculation: $12\frac{1}{4} \div \frac{3}{4} = 16\frac{1}{3}$." Very few sixth graders can construct an appropriate problem for this and explain what the answer would mean, leading to the sad conclusion that much of the time spent learning *how* to divide fractions is wasted — they don't know the most crucial part: *when* to divide fractions.

All such questions, and real-life problems in general, become penetrable and do-able if the student carries a clear set of ideas about the meaning of the four operations. Then the "fingerprint" of the required operation (that is, the configuration and conditions necessary for using it in a problem) becomes obvious in any problem. The key to retaining these operational fingerprints in usable form is to learn them as more than *just* symbols. I like to tap the bodily-kinesthetic, spatial, linguistic, and logical-mathematical intelligences to create understanding and retention of the *meaning* of these concepts. (You can tap the personal and musical intelligences also — see below.) So let's consider one operation at a time, beginning with the easiest and working up.

Addition

Addition is the first learned and simplest to grasp. Young children like it partly because it makes things accumulate and grow — so much so that they often inappropriately choose it when another operation is called for!

A strong kinesthetic association with adding is the word **"Smoosh."** The *gesture* for this operation is to *hold the hands apart in front of the body and, as if a wad of a real or imagined something is stuck in each hand, slowly bring them together to form a lump. This is done while saying "add...plus..."* then *"...smoosh". "Then I name the result."* If clay is available, use it for this. Otherwise, suggest using clay or some other goopy stuff in the imagination.

The *fingerprint* of addition is that at least *two amounts in the problem are supposed to be stuck together to get more, which is named.* Go on to discuss that putting the two "smooshees" together will make a *total* that is bigger and that *includes* both of them.

Subtraction

Subtraction is **gestured** by *holding one hand still and open in front of the body while the other hand seemingly grabs a piece of what is in that hand and pulls it away.* The *first* hand is then focused upon because it is the remaining amount that is the answer to the subtraction. The word *"subtract, minus..."* is said before this gesture and **"take-away"** is said during it.

Note that subtracting is *not* separating two numbers, but *removing* one amount from the other and *naming what is left.* Students should also be able to model subtraction symbolized as 23 - 18 with manipulatives, by removing 18 counters from a pile of 23 counters, taking the 18 elsewhere, and grasping the 5 remaining counters as the answer, saying "5."

The *fingerprint* of this kind of subtraction is that *some number must be removed from another, and the amount left over is being sought.* It should be pointed out, either in discussion or presentation, that you need a bigger starting number before you can remove something smaller from it, and the bigger starter is always written *first* in a calculation. (Later, when they encounter negative numbers, they can modify this rule; numbers can be in any order.)

A next level of sophistication is the use of subtraction to **compare.** In this situation, the subtracted quantity is *not* removed from the bigger amount. Rather, using counters, the smaller quantity is placed side-by-side with the larger quantity, and the amount of the larger that exceeds the smaller is "the difference" or the amount by which it is greater. It is computed by subtracting larger minus smaller.

Let me give you an example of a problem with the comparison fingerprint of subtraction: If there are 14 apples and six oranges on the table, we cannot "take away" the oranges from the apples. But, by subtracting their numbers, we can state that there are eight more apples than oranges. The *fingerprint* is simply that *two amounts are being compared, and the "extra" that one amount has beyond the other must be found.* The **gesture** for this kind of subtraction is *to hold two hands horizontal pointing towards each other, but at different heights out front of the eyes, as if comparing two heights.*

Multiplication

Multiplication brings a new level of subtlety. The word itself is from the Latin *multi-*, meaning many. So multiplication means to "make many" or, to coin a term, to "many-ize." For example, the meaning of "five times 12" is exactly that: 12 happening five times. That's why it starts off "five times..." — to indicate that something is happening five times. That something is 12. Therefore, the *first* number in a multiplication tells how many times the *second* number is to be made many. To put it even another way, "five many-izes the 12."

Portraying this example with manipulatives, 12 counters are placed in a tight row. Then four more 12-rows are placed near it (not touching it) to make five in all. The total of the counters in this array becomes the "product" of the **many-ization**.

This means that the first number and second number of a multiplication play different roles. A visual metaphor for multiplication that most students are familiar with is the **photocopier**. If I place an array of 12 objects as a master on the copier glass, then push a five button to set it for five copies, I get five 12s coming out for a total *product*-ion of 60. I like to say "What's the image for multiplication?" To which they answer with gusto, "A photocopier!"

Adding is smooshing — subtracting is taking away — multiplication is many-izing products — division is cutting or packing.

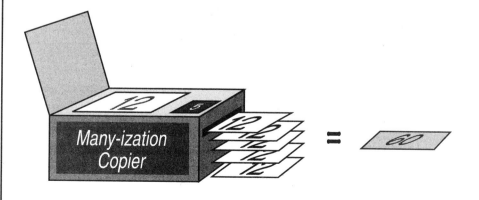

The *gesture* for multiplication is to *extend the left hand palm up in front of the body while the right hand, palm down, makes several motions off of the left hand, resembling duplicates of the left hand being made.* The student says, "Multiplying, many-izing," while doing this. The *fingerprint* to look for in a multiplication problem is *two numbers, one meaning some kind of amount and the other counting copies of that amount.*

Let's take an example: "I plan to drive for four and three fourths hours at an average speed of 52 miles per hour. How far will I get?" Most would probably arrive at multiplication in an attempt to get a big number. But which multiplication most accurately represents the necessary computation, 52 x $4\frac{3}{4}$ or $4\frac{3}{4}$ x 52? The latter one does because every hour I travel 52 miles, so I manage to drive $4\frac{3}{4}$ copies of 52 miles in all. The 52 is happening $4\frac{3}{4}$ times.

A common objection to this approach is, "Yeah, but either order produces the same answer. That's commutativity! So why make a big deal about the order of the two numbers?" There are *three* reasons to make a big deal of it. The *first* is that four times 12 and twelve times 4 can be *very* different processes in life. Consider crossing a shallow river with twelve 4-foot jumps on rocks versus crossing it with four 12 -foot jumps! What's the difference – either way the river is 60 feet wide! The difference is that one way is physically possible and one is not. The only realistic multiplication going on here is 12 x 4, not 4 x 12.

The second reason for the big deal about multiplication order is that a learner needs to be able to *feel* multiplication going on in the problem so that there is never a chance that another operation will be chosen. It can be *felt* if the mathematical symbols exactly correspond to the parts of the problem. In the river situation, you will never feel 12 being repeated four times, but you *will* feel four being repeated 12 times. And thus, 12 x 4 = 48 is the sentence which reflects understanding.

The third reason for stating multiplications in the correct order is that without it, students encountering fraction and decimal multiplication will make little real sense of something like .7 x .34 = .238. Where could such a calculation occur in real life? Why is the answer smaller than either number? How could such an answer have been anticipated so that if my calculator got 2.38 instead, an error would be detected?

Here's where we go into the extra feature of the many-ization copier. Suppose I place a 12 on the copier glass. There are buttons on this copier that allow me to punch $\frac{1}{2}$ for the *number of copies.* What comes out is a $\frac{1}{2}$ copy of 12, or 6. Thus, $\frac{1}{2}$ x 12 = 6 (i.e., 12 happening only $\frac{1}{2}$ (of a) time is really 6. (The x-sign is best pronounced "time" here rather than "times" to secure the meaning.)

Twelve isn't happening even *one* time, only a *half* a time, thus we get 6. That's why multiplication by fractions or decimals shrinks the answer. That's also why you may have learned that "times" and "of" are interchangeable when they occur after a fraction, decimal, or percent: $\frac{1}{2}$ *times* 12 gives $\frac{1}{2}$ *of* 12 since 12 is happening only $\frac{1}{2}$ time. The *fingerprint* of this kind of multiplication is *that we need to find a <u>part</u> of a number or of another fraction (or decimal)*. The hand *gesture* is *to do the same multiplication hand movement described above but with only two fingers and a single stroke, indicating a partial copy*. The simple task, then, is to remember that finding a fraction (or decimal part) *of a* quantity is accomplished by multiplying the fraction (or decimal) *times* that quantity.

Now things like $.7 \times .34 = .238$ can make sense. We are finding a part of .34. Here, .34 is happening .7 of a time, not a whole time. Most of .34 is happening, but not all of it. Thus .238 is a reasonable answer. How can $3\frac{3}{4} \times 9\frac{2}{5}$ be estimated? This is a bit more than 9 happening roughly 4 times, for an answer of about 36. ($35\frac{1}{4}$ is the correct answer).

*Division
as cutting
occurs when
both numbers have
different units.
Packing
occurs when
the units
are alike.*

Division

Then what could be left for **division** to do? Plenty. Let's start with this request: "Make a cloud of 15 dots on a piece of paper. Circle groups of dots to prove that $15 \div 3 = 5$." Some tend to circle groups of three dots until the dots run out. Others tend to circle three groups of five dots. In fact both are right. Dividing or partitioning a group of 15 objects using the number three has two quite different meanings that show up in two quite different kinds of story problems.

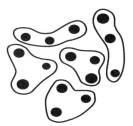

Packing (unpacking)

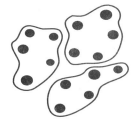

Cutting (dealing)

The left illustration shows the circling of three dots at a time until the 15 run out. Then the *number of groups* resulting is counted: 5. This is akin to seeing how many threes will *fit into* 15 (or can be removed from it). We were taught to say, "Three 'guzinta' 15 how many times?" I call this the **"packing"** interpretation of division, but sometimes I find "unpacking" language handy also.

The right illustration takes the tack of partitioning the group of 15 with a three (i.e., **cutting** the group into three *equal* parts),

with a 3-cutter, then announcing *what number of dots is in each part*: 5. I call this the **"cutting"** (dealing) interpretation of division. Note that you don't have to know the answer is five in advance of doing the cutting. In some card games a deck of 52 is dealt to four players (52 ÷ 4) and each gets 13 cards. The 52 cards are cut by the four players. You could cut 15 with three if you "deal the dots out" to three positions until they run out.

Sample packing problem: "Due to a bladder condition, Jean must stop every 50 miles to use the restroom on her 650-mile drive. How many stops must she make, including the one at her destination?" (How many 50-*mile* chunks will pack into — or unpack from — 650 *miles*? Ans: 13 **stops**) NOTE: In packing, the units of both given numbers are the <u>same</u>, *miles*.

Sample cutting problem: "A highway department wishes to place 50 speed-limit signs along a treacherous 650-mile stretch of road. How far, on the average, should it be between each sign?" (650 miles is to be cut into 50 chunks. Or picture the 650 *miles* being dealt out to 50 *signs*. How many miles end up in each chunk? Ans: 13 **miles**) NOTE: In cutting, the units of the given numbers are <u>different</u>, *signs* and *miles*. That is, you can **cut** 650 *miles* with 50 *signs* (deal them to 50 signs).

While I would not recommend getting completely hung up on this distinction in every division problem, it is extremely handy for helping students know exactly what is going on in a problem — to *feel* division going on. Also, in later fraction and decimal work it will greatly aid understanding.

For instance, the denominator, 5, of the fraction $\frac{2}{5}$ is a *cutter* that carves a whole into five equal pieces and names their size. (See figure.) So, to find $\frac{2}{5}$ of 20, we can divide (*cut*) 20 by 5 producing five 4-pieces. The numerator 2 *chooses* two copies of these 4-pieces, resulting in 8 as the specified part of 20. The numerator is thus acting as a *multiplier* while the denominator is acting as a *divisor*.

Another example of the power of this imagery is to handle the question asked earlier, i.e., making a problem in which $12\frac{1}{4} \div \frac{3}{4} = 16\frac{1}{3}$ would be the calculation. For fraction division, use the *packing* meaning of division. How many $\frac{3}{4}$ can *pack* into $12\frac{1}{4}$? Since 1 can pack into $12\frac{1}{4}$ exactly $12\frac{1}{4}$ times, we expect $\frac{3}{4}$ to pack in *more* because it's smaller than 1. The answer says that $16\frac{1}{3}$ of these $\frac{3}{4}$ amounts can fit into $12\frac{1}{4}$.

Now, the problem:
"Suppose you have to use some $\frac{3}{4}$"-wide transparent tape strips side-by-side to cover a $12\frac{1}{4}$ "-wide piece of cardstock. How many such widths will you need?" Ans: $16\frac{1}{3}$ widths, where that last $\frac{1}{3}$ width means $\frac{1}{3}$ *of* $\frac{3}{4}$" or $\frac{1}{3}$ x $\frac{3}{4} = \frac{1}{4}$" of the full $\frac{3}{4}$" width of the tape.

Even a calculator knows <u>how</u> to divide; only a human can know <u>when</u> to divide in a real-life situation.

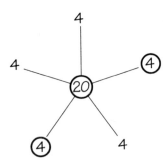

5 cuts, then 2 chooses:
so $\frac{2}{5}$ of 20 = 8

The *gestures* for division are *the right hand pushing into an imaginary bag held by the left hand for* **packing**, *and the right hand chopping along the left arm as if producing slices from it for* **cutting**. The *fingerprint* of division is that *a number is given in the problem that needs to be partitioned or broken up and another number is given that either must fit inside it several times or cut it into several pieces.*

More intelligences

If you want to also involve the **intra- and interpersonal intelligences** up to about the fourth grade, use the Waldorf primary school method of personalizing each of the operations with a name and style, like Arthur Add, Sally Subtract, Mickey Multiply, and Darlene Divide. Each gets into story situations where they want to do the thing they do best. For instance, in the cafeteria Mickey notices five students at each table and 14 tables. He proceeds to calculate how many students there are. He also enjoys arranging several copies of a pair of dessert cookies. As for Darlene, she is seen as being very fair in her group of friends because she always splits up the popcorn in exactly equal portions. If you hand her a piece of cloth, she can rapidly check how many of those pieces could be removed from a bolt of the cloth. And Sally likes to take but has trouble with giving, etc. Students can dress and role-play these operational personae, even coming up with playlets about them.

The **musical-rhythmic intelligence**, too, can come aboard if you make a distinctive jingle or characteristic few notes on an instrument to go with each operation. If students are stuck about choosing the right operation, or tending to choose the wrong one, a melodic hint, like the division tune, will shift their approach while still leaving them the challenge of justifying the use of that operation in the problem.

There you have it, a way to get students linguistically, visually-spatially, kinesthetically and logically involved in the meanings of the four operations, with options for the personal and musical-rhythmic intelligences as well. In my opinion, the sooner the students have internalized the real **meanings** of the operations, not just the rote mechanics of how to do them, and the sooner they can locate the characteristic fingerprints of the operations in realistic problems and investigations, the sooner will they relate math to their world and make a large jump forward in their mathematical reasoning skills.

Sharpening the Math Facts

STARTING AND HONING THE TABLES WITH MI

I remember Sheila, a sociable (interpersonally skilled) young lady in eighth grade, who was receiving low math grades regularly. She had learned her math concepts only mechanically, but her assets included fairly strong linguistic and interpersonal intelligences. She was a finger-counter and shoulder-shrugger when trying to recall her addition and multiplication tables, which meant that her subtraction and division facts were very slow also. She felt embarrassed that she still didn't have these lower-grade skills mastered. Meanwhile, her parents were getting discouraged about her math progress and saw her college hopes fading. So they took a three-hour car ride to see if there was anything *I* could do for her.

After ascertaining that many facts were weak, I met Sheila on the *intrapersonal* front first. I pointed out to her that though these little facts seemed colorless and far less interesting than other parts of her life, her lack of rapid recall of them was affecting her feelings about herself — damaging her self-esteem. It was also causing her work to be slower and less accurate, thus undermining her confidence and generally making academic life more of a downer.

To reinforce awareness of how unmemorized math facts were slowing her down, I asked Sheila to multiply a three-digit decimal by a two-digit decimal on paper. Whenever she paused for lack of knowing a math fact, I would instantly jump in and act as her "inner memory-voice" by whispering into her ear what the answer was. She would continue her written calculation without missing a beat. Then, after asking her to reduce three fractions, I did a similar "act" as she searched numerators and denominators for like factors. She saw that what were usually tedious chores became rapid manipulations when she had an inner memory-voice that knew math tables.

Then I went to the *interpersonal* front. I pointed out that her memory for numbers wasn't defective — she had memorized several of her friends' phone numbers almost as soon as she heard them. I predicted that her friends would be pleased when her math skills improved (because she would be generally happier) and that it wouldn't have an adverse effect on her popularity with boys to know math better.

Now that her motivation was higher I could enter the visual-spatial, kinesthetic, and logical-mathematical realms to take her through some routines (described below) for addition. After a break, I showed her some patterns in the times tables and suggested some memorization techniques for the "stubborn" facts. I saw her a week later and she knew all of both tables by heart, something that had eluded her in years of math education. As a result, we had time to go on to other conceptual topics she'd been fuzzy about.

This anecdote reflects one of the most common challenges for teachers of the middle grades — getting students to learn (and stay sharp with) the *math facts* for addition, subtraction, multiplication, and division. For lack of time, many teachers simply try to ignore the gaps their students have in this area and plug along with instruction, hoping speed and accuracy will improve automatically. Their students find the table values they need by counting on fingers, counting through multiples, and consulting printed tables, but their proficiency with numbers remains halting and stumbling.

Other teachers (and most texts) spend valuable class time in the first part of fourth and fifth grade re-teaching the tables supposedly learned in the previous year(s). They generally use only one or two teaching techniques: practicing timed pages using computer drill software, or assigning flash cards. Meanwhile the class is split among those students who already know their facts, the ones who are a bit shaky, and those who will continue, despite all review, knowing the facts only in a hit-or-miss fashion, even into high school and adult life (and who will be burdened with an inward stigma they've placed on themselves because of it).

The biggest mistake we can make in this situation is to assume that all of our students will learn the tables the same way we did or the way a textbook shows them. In my experience, there are almost as many individual scenarios for how kids learn the tables as there are students. Besides differences in mental development curves, there are different intelligences being used, with different motivations, different parental involvement or modeling, and different learning styles, not to mention the different ways students choose to demonstrate their mastery. Furthermore, some of your students may be harboring trauma, boredom, or just plain turn-off, engendered by previous learning experiences or by desperate reactions of teachers and parents to past failures.

You may recall my anecdotes in the introductory chapter of this book, telling how one student learned rapidly from a music tape made by her dad, another by making colorful pictures out of each math fact, and yet another by jumping rope. I'll expand on these methods of memorization later. First, however, I want to outline a systematic approach that I use with great success. Many individual students have made a complete turn-around in a few days because of it.

The math tables are the bottom bricks in a whole structure of higher math skills

Teaching the addition facts

(Read through this even if you plan to teach only multiplication.)

Drawing on the mathematical-logical, visual-spatial, inter-personal, and linguistic intelligences, this method leads many students away from finger-counting once and for all so that, in a matter of hours or days, they'll be able to do rapid addition facts.

As I illustrated in my introductory anecdote, it's wise to start with a motivating pep-talk and demonstration of the advantages of knowing the addition facts "cold" — with the fingers still. It's ten times easier if students *want* to be able to recall the facts rapidly.

After that I launch into why *ten* is the "big shot" — the queen, king, or "cool mover" of the number system (depending on which image is a grabber). It even helps to get in a bit of a "cool" jazzy mood and convey these ideas in a secretive or storytelling voice. "10 is *it*, it's cool. How do we know 10 is the most important number? If you want to know who's important in a country, check the *coins*. You always see presidents and rulers on them. To see who's important in Numberland, check the numbers. See who's engraved on them.

"Take *seventeen*, for example. Let's write it *seven-teen*. Now *teen* is a secret code word for *ten*. So *seventeen* means *7 and 10*. That means 10 is engraved on all the teens numbers. Check it out!

"Now take *60*. That's 6-ty. In the olden days, *y* was used in place of *e* in many words because it often makes the same sound as an *e* like in *many* and *baby*, and it stands in for *i* too, as in *fly* and *cry*. Well, the *-ty* in sixty stands for *-te*, and this is another secret code for ten. So *six-ty* means *six tens*. A -ty is a 10 wherever it occurs.

"That means almost every number you say has 10 encoded in it in some way. Can you think of exceptions? What about 100? Well, that's really stamped with 10 because it's *10 10s*. So what about 7? It's one of *ten* symbols that make all the numbers. And eleven and twelve? They're 'oneteen' and 'twoteen,' but because they live so close to the leader-ruler 10, they have the privilege of different names (but their names' ancient roots salute ten by meaning 'one over' and 'two over.')

"OK, so 10 is cool. How does that help to learn addition or multiplication tables? That's easy. If 10 is cool, how do you think 9 feels? One-down, of course, and, though it could choose to be perfectly happy with its size, it won't let go of wanting to become a big-shot 10.

"Check out what happens when 9 meets 7, for example. It sees a possibility and strikes a deal. It says, 'How would you like to hang out with a big-shot 10?' Now 7 is a bit ambitious also and says, 'That would be cool!' Then 9 says, 'Fine, but you'll have to pay a little price. If I can take a one from you, I'll become a 10 (with the title 'teen'), and you'll become a 6. We'll be 6-teen together.' And so they do the exchange."

Have students practice with counters and in their heads what happens when 9 meets various numbers, even 2. They should rapidly arrive at the resulting sum. I call this the **"Hungry-9"** trick. Students who have given up on ever stopping counting on their fingers get very inspired by their new-found speed with this simple trick.

"But what about 8? It's pretty ambitious, too, and would like to be a big-shot 10. How much does it need to do it? Just 2, of course. So the hungry-8 strikes a deal with 5 and says 'Give me 2, and you can hang out with a big shot.' 5 goes for it, gets reduced to 3, and joins the new teen to become '3-teen,' better known as *thir*teen." It's only necessary to practice this **"Hungry-8"** trick on 8 + 2 through 8 + 6. Other tricks will cover the rest.

"Now, even **'Hungry-7'** wants to become a big-shot! But it needs 3 to do so." The only practice students need to do with this one is 7 + 3, 7 + 4, and 7 + 5.

Next it's important that students know all of their **doubles**, like 7 + 7. Most finger-counters already know these by heart. I have seldom found a student who can't remember all the doubles from 1 + 1 to 9 + 9 after a few short rounds of recitation or writing practice.

Knowing the doubles, they can move to the **"Off-the-Doubles"** trick, which is based on easy recall of the doubles. How much is 6 + 7? When students hesitate or start moving their fingers, I say "How much is 6 + 6?" "12" is immediate, so I quickly say, "Then how much is 6 + 7?" That's usually enough to start the trick, but if not, a simple discussion or use of manipulatives will convince them that 6 + 7 is just one more than 6 + 6. This approach leads to the rapid learning of 2 + 3, 3 + 4, 4 + 5, 5 + 6, 6 + 7, and 7 + 8.

When these become known well, all that remains is to clean up the **small facts**, the ones that come out less than or equal to ten. The only remaining troublesome ones are:

2 + 4, 2 + 5, 2 + 6, 2 + 7   3 + 5, 3 + 6   4 + 6.

These can become rapid in a variety of ways. Here are a few suggestions. Counting *odds* and counting *evens* to 20 can pave the way for memorizing the first 2s group (since every answer is just a 2-jump beyond the previous one). Count evens and odds rhythmically with clapping, bouncing a ball, or jumping rope.

Noting the 3-6-9 positions on a clock face and counting by 3s should cement the 3 + 6 fact. The 3 + 5 and 4 + 6 can be seen to be a kind of off-the-double variation, where adjusting one up by 1 and the other down by 1 results in a double. That is, 3 + 5 = 4 + 4 = 8, and 4 + 6 = 5 + 5 = 10. Some practice with manipulatives and visualizing this will usually lead to automatic recall of the answer.

Rapid recall of math tables hastens calculation, highlights number relationships, and aids mental math, estimation, and problem solving.

These various facts need to be practiced daily for a while, first grouped by trick, then randomly. A good source of random questions is a set of playing cards with its royalty and aces removed. A student can shuffle them, then remove two cards at a time and state the sum. Students can take turns around a group doing this.

To speed things a bit, kids can play "war" with the cards. Deal out the deck completely to the players, who stack their decks face-down before them. A "play" consists of each player taking two cards from their decks and turning them face up, then saying the sum. (An additional interesting rule: If a player says a wrong sum, the first other player to say the right sum automatically gets the first player's cards.) The highest-sum player takes all cards played. In the event of a tie, a "war" is declared, and the tied players turn over the next pair of cards from their decks. The winner of that dual takes *all* the played cards. When a player's deck runs out, his or her winnings are shuffled and play resumes. When a pre-arranged time-period runs out, or when any player has one or no cards left, the game is over, and the player with the most cards wins.

Mild timing pressure can be put on by seeing how many pairs of cards can be drawn and summed in exactly *one minute*. A standard can be set by laying 15 counters in a line on the table. Note when the second hand on a clock is at a good starting place (like straight up). Each time a pair of cards is drawn and summed correctly a counter is hastily pushed aside from the line. The goal is to move all 15 counters in a minute. If a student can do this one-minute criterion three times at a sitting, I consider the student fairly proficient. An alternate and very engaging producer of random additions arises from pasting numbers onto, or writing on, two regular, wooden or foam *dice. (See the Resource Bibliography: Manipulatives — Wooden or foam cubes.)* To catch the most important facts, mark the dice 3, 4, 5, 6, 7, 8 and 4, 5, 6, 7, 8, 9.

The **15-in-a-minute criterion** can be achieved by a whole class using cooperative learning, especially if a majority of the class has not yet mastered the tables. In groups they try to raise their adds-per-minute only 15 minutes daily for a few days. Arrange the groups so that there is a range of proficiency in each group. Offer a reward to any group that has all of its members at proficiency; offer another reward for the whole class when it is at proficiency.

Encourage struggling students — whether they're in a class that's working on the tables or one of a few that still don't know them — to work on this at home with a selected card deck or with dice. An offer of some kind of reward from family or teacher can be helpful here (a non-candy variety is preferred). Enlist the student's family, as well as other students, in the challenge. Make sure family helpers know the various adding tricks so they can cue when a troublesome fact comes up in the cards (or dice). For example, when 6 + 7 comes up, and there is hesitation, say, "Off-the-double" or just, "Double."

Teaching multiplication facts

(Read through the addition section first to pick up many general pointers and to get the right terminology and mind set.)

Most students feel that there are at least a hundred facts to learn with brute memorization when the reality is that there are barely 15 that require *real* memorization effort or technique. Here are the reasons why:

➤ Almost every times fact is given *twice* in the table: 3 x 4 *and* 4 x 3. My experience is that when one of them is learned solidly, the other is *usually* learned almost as an afterthought. Sometimes a turned-around one can be sticky.

➤ The **1's table** is already known.

➤ The **2's table** follows rapidly from knowing the doubles in adding (i.e., 6 + 6 = 12 can be translated easily for students to 2 x 6 = 12), especially when they are clear about the meaning of multiplication. *(Make sure they are! See Chapter 9 on "Power-packing the Operations with MI.")*

➤ The **10's table** is a piece of cake — just place zero on your multiplier. (Make sure to discuss why this works, using manipulatives, etc.)

➤ The **11's table** is cake also — just double the multiplier's digit: 7 becomes 77. (Be sure to demonstrate that this works due to 11's proximity to 10 — 11 = 10 + 1)

➤ Much of the **12's table** can be postponed until later since it isn't used a lot in basic computations. The first four 12s have a pattern that can be taught immediately. The pattern of 1-2, 2-4, 3-6 and 4-8 is easy to see — just double the multiplier and make that the second digit. (The 5 x 12 will be rapidly learned with the 5's trick.)

(Later, after all the lower tables are learned, you can point out that 6, 7, 8, and 9 times 12 has the pattern of raising the multiplier by 1 and making 2, 4, 6, or 8 the second digit. After this is practiced, teach that in 11 x 12, the 11 tries its trick by repeating 12 twice, like 1212, but that's too long of a number, so the middle two digits add together and we have 132. Then 12 x 12 echoes the first four 12-facts

➤ "Because 10 is the big shot, some numbers get more of a chance to do smooth tricks in the times tables than others. We encountered 10, 11 (just one more than 10), and 12 (two more than 10), above. But 9 (just one less than ten), and 5 (half of ten) do great times table tricks."

➤ The **9's table** can be taught very early to instill confidence that the tables aren't so bad. Nine can do its trick because it is so close to 10. Show them this pattern of the nines answers and invite observations:

X	1	2	3	4	5	6	7	8	9	10
	09	18	27	36	45	54	63	72	81	90

They will notice that each 9 answer has a reverse counterpart. Some may notice that the *sum of the digits* in every answer is 9. A few may notice that the multiplier number minus 1 gives the first digit of the answer, e.g., **6 x 9** starts with **5**, and **8 x 9** starts with **7**.

This gives a key to mentally finding the 9's answers. For instance, 7 x 9 must start with 6 and the second digit of the answer must, with 6, make 9. It can only be a 3. Voila! 63 and 6 + 3 = 9. A little practice of the adding combinations that make 9 and of this trick can make most students proficient in the 9's table, asked randomly, in about 15 minutes.

➤ Finally, the **5's table** is simple because 5 nicely relates to 10. Kids easily know that two 5s make 10. A next question can be, "Then what do *four* 5s make?" Aha, *two* 10s, or 20. "How about *six* 5s?" Must be three 10s, or 30. Similarly, *eight* 5s make *four* 10s, i.e., fo(u)r-ty or forty.

It can easily be shown with manipulatives, such as interlocking cubes or Cuisenairre rods (*See the Resource Bibliography: Manipulatives.*), that a bunch of 5s can be recombined to make half as many 10s. For instance, twelve 5s would make only six tens, or 60 and 14 5's should make 70.

This only leaves the weird fives 3 x 5, 5 x 5, and 7 x 5 to be learned. A little manipulative work reveals that there is an extra 5 on each answer. Or, putting it another way, 7 x 5 will yield only $3\frac{1}{2}$ tens, or 35. Similarly, 3 x 5 and 5 x 5 yield $1\frac{1}{2}$ and $2\frac{1}{2}$ 10s respectively. Most students can make the jump to this trick rapidly, and virtually all, by practicing for a while with manipulatives, can do it.

Many kids don't know ten is the most pivotal number in our system despite their early place value work.

After the systematic patterns on the previous page have been practiced as independent tricks and each one is well known, they can be mixed in practice by making and using **two dice**. Die A has 2, 5, 9, 9, 10, and 11, for instance, on its faces (to create more 9-practice). Die B has 4, 5, 6, 7, 8, 9 on its faces. Students practice throwing the dice and multiplying the faces. (A Die C with 3, 4, 5, 6, 7, 8 on its faces can replace B, to bring 3 into these tables.)

Once these are fluent (and this can happen in well less than a week), it's time to master the **4-tables**. The only 4s left to conquer after the previous work are 4 x 4, 4 x 6, 4 x 7, and 4 x 8. (Note that the *reverses* of each, like 6 x 4, are math facts too, but learning them solidly one way paves the way to easy absorption of the reverses.)

The number-sense way to learn 4s is the "Double-Double Trick." So 4 x 6 is just 2 x 6 doubled. The student says internally, "6-12-24" as the 6 double-doubles. Similarly, 4 double-doubled is 4-8-16; 7 double-doubled is 7-14-28.

Students can master these quite rapidly. The hardest double-double is 8-16-32. Prep their minds by discussing 15 + 15 = 30. It is a very important *clock fact* and it's modeled easily with two 10-cube and two 5-cube stacks. Now 16 = 15 + 1 so 16 doubled makes 32.

Now the 4s are cruising. Reinforce them: students in pairs throw the C-die above and practice the 4-multiple of each throw. **Next come the 3s.** Though a little harder to master, only four need special attention: 3 x 3 (often mastered already), 3 x 6, 3 x 7, and 3 x 8.

Introduce the "Double-And-One-More" trick, e.g., for 3 x 6, double 6 to 12 and add 6 again. Or for 3 x 8, double 8 to 16 and add 8 (jump 4 to 20 then 4 more to 24). 3 x 7 is tougher because adding 14 and 7 to get 21 is not mentally simple; discuss and practice this 14 + 6 + 1 jump.

It's also helpful to "prime the 3-pump" first, or simultaneously with, learning the trick. Here are some interesting and motivating MI ways to build familiarity with the 3- (and 4-) sequences.

1) Discuss and explore the well-know divisibility fact that every 3-answer has a 3, 6 or 9 digit-sum.

2) Build familiarity with the 3-multiples (and 4-multiples) by having a lot of fun with Activity 5, "Clock Pictures," in the back of this book.

3) In the multiple-intelligences groove, let kids practice the 3-multiples (forward and even backwards) kinesthetically and rhythmically by reciting them while regularly, steadily dribbling a ball, jumping rope, or bouncing a ball off a wall and catching it. Another method, often used with special-needs youngsters, is to hang a small rubber ball from a string attached to the ceiling.

There is no substitute for short daily practice while learning the math facts (and after learning them also).

The student holds a small board or book that is steadily "bonked" against the swinging ball in rhythm while reciting 3s or 4s.

4) Try singing "3 x 3 is 9, 3 x 4 is 12 ... " to a standard or made-up tune (like " ... dashing through the snow .... ").

6) For the visual-spatial or visual-symbolic learner: 1) "draw a picture with the sequence embedded in it, 2) read the sequence over and over on paper, or 3) write all its numbers *upside-down* (evoking single-pointed attention while engaging the right brain).

A simple **game** for any math-facts reinforcement involves making a grid of boxes (e.g., 12 boxes in a three-by-four arrangement or 20 in a four-by-five arrangement), writing the fact-answers randomly in the boxes (repeats are okay), then players take turns throwing dice and placing their color of counter on the answer in the grid. The first player with three (or the rule can be four) counters in a row, column, or diagonal wins. If transparent colored tokens are used an alternate rule can be that an answer number with one player's red token on it can also be played by the other player's blue token in the attempt to get four in a row. *(For tokens, see Resource Bibliography: Manipulatives.)*

16	32	28	16
24	24	28	28
16	32	32	24

A **solitaire** version of the game above can be structured by drawing two separated cliffs with a row of boxes making a bridge across them. In the boxes are various answers to the selection of facts being practiced. A die or dice are to be thrown a pre-set number of times, like 15. The goal is to place tokens in the answers to questions that come up on the die (or that result from multiplying two dice). If, in 15 throws, an unbroken row of boxes is filled so that someone could get across the bridge between the cliffs, the player wins the solitaire.

By the time all of the facts above are learned, there are only **six facts left** in the below-12 category. These are: 6 x 6, 6 x 7, 6 x 8, 7 x 7, 7 x 8, and 8 x 8. There are many intelligences and techniques to "cement" them. Here are some suggestions:

Help students make **mental associations**. Here are some:

- 6 is a poet, or better a *half-poet*. Here's why: clap rhythmically and recite to this rhythm: "6 x 4 is 24, 6 x 6 is 36, 6 x 8 is 48, but...6 x 7 is 42." That is, most rhyme but 6 x 7 is the non-rhyme exception. YET, even it has the "half"-trait, i.e., one digit of the answer is half the other (but note these reverse in 42).

- Write the 7's with loooong tops: " $\diagup$ x $\diagup$ in a line makes 49."

- Write these numbers in order: 5 6 7 8. Then punctuate it this way: 56 = 7 x 8. This fact is the 5, 6, 7, 8 fact.

- "I ate and I ate till I was sick on the floor." 8 x 8 = 64 becomes easy to remember with this strong image. Also 8, 6, 4 are descending even numbers.

Other ways to memorize "stubborn" facts

(Note that these techniques will work for other factual information as well. Instructions are given as if they are addressed to your students. Separate indications to you, the teacher alone, are enclosed in a grey box.)

➤ **These are some old-fashioned ways** we all grew up with and they might work for some of you.
  1. Say the stubborn ones over and over. (Verbal-Linguistic)
  2. Write them several times. (Visual-Spatial, Kinesthetic)
  3. Read them over and over, with and without answers, on flashcards. (Visual-Symbolic)

➤ Go through a list of your stubborn math-fact questions. If an answer doesn't come instantly, consult the **calculator**. Seeing the answer on the screen can be a reminder. (Visual-Spatial, Kinesthetic)

➤ Join the fact to a **memorable event**. (Visual-Spatial, Kinesthetic, Interpersonal)
  1. Make it large, i.e., with chalk on a sidewalk; or filling a blackboard or poster.
  2. Put it in an odd place: on a tennis shoe, cast, hat, someone's answering machine, muddy car fender, phony tattoo, a cake, a mirror (with shaving cream), the floor (with tape), or a basketball.
  3. Dramatize: Walk up to someone on the playground and say, "Did you know that ...?" or mail it in a nice card to yourself.

➤ Make and use **flashy flashcards** to learn the more stubborn of your facts. That is, make flashcards visually interesting and vary the visuals from card to card. (Visual-Spatial, Kinesthetic).

The usual black-and-white flashcards leave most kids cold. Their main subliminal message is, "I am a sterile and lifeless fact." Flashy cards create spatial and color associations that the answers can attach to.

1. In making flashy cards, remember to have the full fact (7 x 8 = 56) on one side and just the question (7 x 8 =) on the other side. The *full-fact side* should be the *flashy* side and the *question* side should be in *plain black* numerals.
2. Make the flashy cards from different colors of paper and use non-rectangular shapes.
3. On the flashy side use odd-shaped numerals, color in some loops or curves, squish some numerals flat, make others angular, and stretch out others.
4. On the flashy side use different colors of felt pen or colored pencils to make the number facts. You can outline and fill the insides if you want. Fill in around them with borders or artwork. Numbers can even resemble objects.
5. When practicing, keep the cards turned toward their *plain questions*. To answer a fact question, picture the *flashy side* of the card and its answer, or, if stumped, turn the card over and look.

➤ Use the **strong-memory area** in your field of vision.
1. Determine the location of your visual strong-memory area.

This area has been used *very* successfully for teaching spelling, by the way. Here's what it's about. Some techniques of Neuro-Linguistic Programming (NLP) use the fact that our eyes look towards a certain direction (often upper-right) to recall visual information (and other directions for auditory and kinesthetic information). A student can determine this by trying to visualize where the scratches are on her doorknob at home or what the exact pattern on some wallpaper is. The eyes will want to stare in a certain direction for maximum recall. This is the visual strong-memory area.

2. Hold a *flashy flash card* for a stubborn math fact about 18" away in that area, and stare at it for a moment.
3. Close your eyes and visualize that colorful card being still there.
4. Repeat steps 2 and 3 a couple more times.
5. Remove the card and stare toward the strong-memory area again. "See" the card there with its flashy fact on it.
6. After doing this with each stubborn fact, go back through the plain sides of the flashy cards, asking yourself the facts. If one is sticky, look toward the strong-memory area and "see" in the air the flashy fact with its answer.

Don't be afraid to suggest silly associations for remembering a fact. Silliness is one of the best tags for memory.

➤ Write a stubborn math-fact three times: **upside-down, opposite-handed, and backwards**. This engages the right side of the brain as well as the left. (Visual-spatial, Kinesthetic)

➤ Walking, moving rhythmically, bouncing a ball, or jumping gently on a small trampoline while saying (or chanting) the stubborn facts **in rhythm to your movements** will aid memory. (Musical-Rhythmic)

➤ Place some flashy cards for a few stubborn math facts **near your pillow** to look at before falling asleep and after waking up.

> The twilight zone between awake and asleep when you sometimes jerk as you nod is called the hypnogogic state. Just before you're awake you're in the "hypnopompic" state. You're very impressionable at those times, and experiences can imprint on your brain. Have you ever awakened with a song on the radio or in your mind and end up humming it all day? Many have. Hopefully your student will be humming math facts all day after this!

There you have it — a smorgasbord of tools to make the "facts problem" a thing of the past. Undoubtedly there are many more possible ways to remember the facts that you will be provoked to try after you start with the general ones I've given. You'll realize that memory devices can be evolved from your students' unique interests in baseball cards, ballet, video, horseback riding, hopscotch, skateboarding, swimming, marimba, poetry, internet chat, monkey bars, and face-painting.

The main points about math facts I hope you take with you are these:

• The math facts, while well below the more complex math thinking you want to bring students to, can hang up students' self-esteem, computational speed and math progress.

• Learning the math facts doesn't have to be a drudge; it can be rewarding and fun.

• Math facts can be learned very rapidly and permanently (with some continuing maintenance) if the right channels are tapped.

• The math facts have pattern and order in them. They can be taught systematically, and you can tap your students' interests and intelligences for their memorization.

Section B:

14 Activities (over 90 sub-activities) for the Student

An Intro to the Activities

MEETING MATH MOLE AND USING ACTIVITIES

Before you hand out copies of any of the following Activity masters, I encourage you to engage your students in a lively discussion of moles. (Yes, moles!) According to my dictionary, a mole is "a small insectivorous mammal living chiefly underground and having velvety fur, very small eyes, and strong fossorial forefeet." Moles are famous for their diligent tunneling and subterranean exploration.

But there's plenty more to know about moles. Pool the knowledge of everyone in class, then get your naturalist-intelligent students to supplement the information with a look in the encyclopedia or any animal book. Many of the moles' characteristics will surface: their keen sense of smell, their way of operating in complex tunnels beneath the surface, their poor sense of sight linked with a phenomenal sense of direction, and their high level of sensitivity, even to the electrical fields around worms! Yes, they love worms! Then there are the physical characteristics — the pointed snout, hairy body, and strong front digging claws.

Many of these characteristics inspired the creatures that you will find cavorting in the pages of this book. One is the "Math Mole," a morphed mole demonstrating part-human and part-mole characteristics. The other is "Wily Worm," a combination of book worm, composter of ideas, commentator, and juicy prey of Math Mole. Here's what I want to tell you and your students about them. Please summarize this for, or read this aloud to, the students. Showing a few pictures of him from this book on the overhead will add interest.

My name is Mark Wahl; I'm the Math Man, and I designed the Activities in the book your teacher will use to help you think in new ways using all of your intelligences. One day, not long ago, I discovered a large mole-like creature as it was burrowing through the complexities of a math problem. I watched quietly as it stood by its big mound of used-up ideas, collecting new data and combining numbers. I liked this mole immediately because it never made a mountain out of a molehill — it just dug in and got the job done. After a while, I cleared my throat to let it know I was watching, and then I asked, "Who are you?"

I became sure "it" was a "he" as he answered "Math Mole." I realized that he had managed to partly morph himself into a humanoid while keeping the best of what moles have to offer.

I noticed that he, like me, improved his poor eyesight with glasses, but, no matter — the real secret of his success is his ability to <u>sniff</u> a solution from a distance. He knows that just learning the mechanical procedures of mathematics, all the rules you follow to get those pages of drill done, is not enough. The Math Mole likes to really dive in there; he loves to explore all the passageways, get his paws dirty, go after subtle leads (always keeping his sense of purpose and direction), then chomp into real results at the end. The Math Mole respects his teachers who taught him that one must be willing to get beneath the surface of things, even to make wrong turns, and to try and try again.

"Math Mole," I said, "How would you like to join the Math Man in developing a set of activities for students?" He exclaimed , "I would really dig that!" And ever since, that Mole and his family have helped me to unearth new ideas, to scoop out new mathematical channels, to grub around in numbers, and to make math activities more fun by acting out different roles.

One of the worms he chased to the surface has also joined in on the act. This one I call "Wily Worm" because it always cleverly figures ways to avoid getting stuck in unworkable plans and it tricks you into seeing things the right way. It must act quickly and keep close to its hole, however, lest Math Mole come too near.

I'll end with some advice Math Mole gave me — you might be able to use it as you do the Activities: "Set your mind on your goal so strongly that you can almost smell it," he says. "Develop a plan, then dig in. Don't worry if things seem messy at first, because

messing around in the numbers is good while you sniff out the possibilities. Then undermine all the obstacles as you stalk your solution. The prey will be very tasty."

Let Math Mole, his family, and Wily Worm encourage you, inform you, and entertain you as you work. Keep their ways and advice in mind when you feel stopped by a rock in your path of thought.

You may want to just leaf through the Activities and see what strikes you as interesting for your students, then try it out. The "Teacher Notes" will make the material self-explanatory.

The Activities are not intended to be done in the order they are presented, but only in the order of your needs and interests. You can consult the **Finder Key for Activities on pp 86-7** (with dark stripes on each edge) to search for particular math skills, levels, or even the intelligences you want to emphasize. *Note: Run off a page or two on the photo- copier for your students. Sometimes you may want to cut-and-paste pieces from pages to create and copy a custom activity handout.*

More about the Activities and their uses

The Activities in **Math for Humans** have multiple uses for the teacher or homeschooling parent. They supplement, further develop, and broaden skills being taught in the classroom. They hone reasoning and exploratory skills, thus supplementing what may be more drill-oriented textbook approaches. They can serve as changes of pace from the usual classroom fare — even as rewards, say, each Friday, after a week of productive but more routine work. They will train you, the teacher, in MI teaching!

Since these Activities help generate a more positive math attitude, you might want to use them more frequently in the beginning of a school year. They'll help you get a sense of the abilities of the students. At the same time, students will be broadening their interests in math.

The Activities are real motivators for learning math skills — they give reasons for learning them. I know a homeschooling parent whose son was increasingly resistant to math books and rote practice. So, for math "lessons," she let him choose activities from an earlier version of this collection. He got so turned on that he wanted to acquire the skills needed to work through all the activities, and then he wanted to continue learning more skills beyond them! Another homeschooling parent alternated those activities with daily lessons from a well-known and very sequentially-organized book.

Mother and daughter both liked how the two approaches complemented each other.

The Activities in this book draw from a range of skills normally found in curricula for grades 3+ through 8, broadening and applying them in new contexts. A range of math processing skills found in the NCTM Curriculum *Standards* (see Chapter 2) is also tapped. And, of course, the Multiple Intelligences are addressed and used throughout.

Each Activity is organized around one topic or theme, then divided into several sub-Activities. Most sub-Activities can stand on their own, but some must follow after earlier ones or, at least after the introduction to the main Activity. This gives teachers flexibility with topics as well as with specific projects for supplementing classroom curricula.

 Watch for the worm in the light bulb, indicating activities that are **more challenging** and that **extend** the activity more for the eager student. They can be skipped, done later, or assigned as extra work for those finished early.

Following each Activity with its several sub-Activities are one to five **Teacher Notes** pages that specify:

➢ **Intelligences emphasized** by the Activities and sub-Activities;

➢ **Math skills and concepts** introduced or practiced in the Activity;

➢ **Materials** needed for the various sub-Activities;

➢ General **Levels of Challenge** posed by the Activities;

➢ An **Overview** of the Activity to orient the teacher;

➢ An **Introduction to the Class** that gives tips on how to present the Activities to the students and how to pave the way to some of the trickier concepts;

➢ **Answers** to all the exercises.

➢ Suggested **Extensions** of the Activity to other intelligences and to other math levels or areas.

You can use these Activities as:

➢ a change of pace;

➢ lead-ins that create interest in concepts to be developed by your curriculum;

➢ a way to bring in a wider range of math thinking;

➢ a motivator for those students who have issues with math;

➢ "food" for those who are hungry for additional math challenge or stimulation;

> a change of "channel," especially for those whose particular intelligence "channels" don't get tuned in by most teaching methods;

> a way to engage parents with students on meaningful home assignments;

> a way to observe and evaluate students' math process skills in action (See Chapter 8, "Assessment Ideas");

> a way for students to keep mathematically active in a homeschool setting as parents are searching for new curricula and new rhythms for their children, and

> material for a parent-child "Math Night" at school, during which parents and their children experience the joy of working together on a meaningful and instructive math project.

Enjoy!

Think of these Activities as "Priming your MI pump" out of which will flow ideas from other resources, and from you.

Finder Key for Activities

ACTIVITY	LEVEL OF CHALLENGE	INTELLEGENCES EMPHASIZED
1. A NUMBER DETECTOR	Easy to moderately challenging	S, BK, I↔P, N
2. SYMMETRY	Easy to moderately challenging	S, BK, I↔P, N
3. CALENDAR CONNECTIONS	Easy to very challenging	S, L, I↔P, BK, M
4. FAIR AND SQUARE	Moderately easy to moderately challenging	S, L, I↔P, BK, M, N
5. CLOCK PICTURES	Moderately easy to moderately challenging	S, L, BK, M, N
6. PEOPLE AND MORE PEOPLE	Moderately challenging	S, I°P, I↔P, L, BK
7. ADDRESSES AND DISTORTIONS	Moderately challenging	S, L, BK, N
8. COUNT ON AFRICA	Moderately easy to challenging	S, I↔P, BK, L, M, N
9. PATHS AND PLANETS	Moderately challenging to challenging	S, BK, I↔P, L, N
10. BE A DATA DETECTIVE	Challenging (with some moderately easy parts)	BK, S, L, I↔P, N
11. AUTOMOBILE$	Challenging to upper intermediate students	I↔P, I°P, BK
12. WATT'S HAPPENING? ELECTRICAL ENERGY	Challenging to upper intermediate students	BK, I↔P, L
13. FROM NUMBER NUMBNESS TO "FEELING" NUMBERS	Moderately challenging to very challenging	I°P, S, L, I↔P
14. GET REALLY SMART: CRACK PROBLEMS WITH <u>ALL</u> YOUR INTELLIGENCES	Moderately challenging to very challenging	ALL

KEY

S = Spatial
L = Linguistic
BK = Body-Kinesthetic
I°P = Intrapersonal
I↔P = Interpersonal
M = Musical
N = Naturalist

MATH SKILLS USED, CONCEPTS BROADENED (by Activity Number)

ACTIVITIES INDEX

FINDER KEY

A NUMBER DETECTOR

Pick a number from 1 to 16 and put it into this machine. Follow the machine as it figures out your number.

If an answer is "yes" take the Y-path.

If an answer is "no" take the N-path.

Teacher Notes

Intelligences Emphasized: Spatial, Kinesthetic, Interpersonal (large group process), Naturalist (sorting, classifying numbers)

Math Skills, Concepts: Mental arithmetic, flow charting, logic, classifying whole-number characteristics (optional choices: primes, evens, odds, divisibility, squares, cubes, triangle numbers)

Materials Needed: Pencil, pen, colored markers, scissors, tape, unlined paper or tagboard, page 1-1 for each student

Level of Challenge: Easy to moderately challenging

Overview: This Activity teaches the skill of using number characteristics for sorting, comparing and contrasting numbers. It can be used to explore simple or more sophisticated whole number concepts ranging from evens to primes and cubes. It's also a nice introduction to the use, power, and logic of flow charting. Students find the Number Detector's seeming uncanny ability to figure out a chosen number surprising and pleasurable.

Introduction to Class: "What devices do you know that *detect* something? (Metal detectors, smoke detectors, etc.) Here's a neat mathematical machine that detects *numbers.* Explore how smart it is by trying the numbers 1 through 16 in it. Then we'll try to make it smarter and we'll even try to *become* a live Detector."

[Answer questions about what Y and N, divisibility, remainders, how to move through a question box in the Detector, etc., as they come up. After students are familiar with the Detector, offer some of the extensions given below.]

Extension: "The Live Detector". (Bodily-Kinesthetic and Interpersonal, for 16 or more students.) Try this after the students are familiar with the workings of the Detector page.

There are 15 question-boxes in the Number Detector. Since you may have more than 15 volunteers and since some of the questions in the machine get far more "action" than others, don't leave it to students to haggle over who will be in the Detector and which question they will be. Let Fate decide. On the Detector page, letter each of the question boxes randomly with letters A to O. Make paper tickets with letters on them from A through enough letters to make one for all the students (go to AA, BB, etc. if more are needed past Z). Each student draws a letter-ticket from a box.

Students with letters A to O consult your Detector page and make that letter's question with large words on a nicely shaped, colorful, sign. Each of the A through O students also marks one hand with a large Y (for yes) and the other with a large N (for no). Students who drew P, Q, R, and S make 16 interestingly-shaped, colorful papers with one of the numbers 1 through 16 on each.

The extra students get the task of brainstorming in groups about how to extend the detector to numbers *beyond* 16 while the other 19 prepare the Detector. (See the Logical-Mathematical Extension below.) Later the extras will get to become numbers going through the Detector and an audience. They may also be able to add workable extensions for higher numbers to the machine after the basic Detector is tried and perfected.

When the 15 signs and 16 numbers are prepared, give the 19 students this **group process problem** to solve (with minimal teacher input):

"Put your sign on your front with tape or a string around your neck. Arrange yourselves in this open area to resemble the Number Detector diagram that each of you has a copy of. Your Y-hand and N-hand will point to the next question or number after you in the Detector.

"For instance, the '2x2x2?' question person Jill must be sure to have her Y-hand ready to point to an '8' on a piece of paper taped to the floor, desk, or chair nearby, and her N-hand ready to point to the '2x1?' question person.

"When you're ready, a number-person will go through the Detector. If he comes to the '2x2x2?' question, Jill will have to decide if his number is 2x2x2 or not, and point, with her correct hand, to the next place the number-person must go. When he lands at a number, he stays there, and a new number-person will come through the detector."

There will be some chaos and group process as the students attempt this feat of arrangement, but invite them to solve as much as possible by themselves. When the Detector is ready, invite a student to make up a number from 1 to 16, put it on a paper and

hold it up as he enters the Detector. He need only follow the pointing of the various question-students until he arrives at a number spot that corresponds to his number.

Extension: (Logical-Mathematical, Bodily-Kinesthetic)
Make the Machine Smarter. Students may devise a further extension to the detector for numbers beyond 16. There are at least <u>three</u> ways to do this:

1) Try to add on questions for just a few more numbers, starting with 17. They do this by putting 17 in the Detector until it fails (by leading to 13).

   The remedy is to insert another question (by pasting a small piece of paper on that part of the Detector) just before the 13, like, "Is the sum of the digits 8?" An N branch goes to 13 and a Y branch goes to a 17 on the small added paper. Each new number requires intervening somewhere in the chart with a question that can lead to it or to an old number position.

2) A different kind of question box that can be used to extend the Detector to more numbers is one with <u>three or four</u> answers. For instance (for numbers over 20):

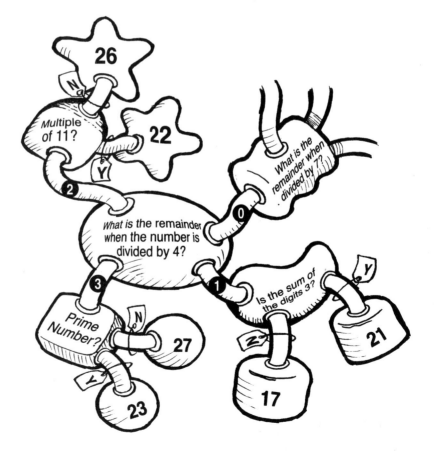

3) Put a question box right at the beginning of the Detector: "Is the number < 17?" The Y is the path into the <u>old Detector</u>, and the N sends the number to a <u>New Detector</u> page geared to, say 17-25, or beyond.

Here are some possible distinguishing characteristics of some of the numbers above 16. There are others. These may start student thought on how to make questions for all the numbers they're working on in the flow-chart for that New Detector section:

17, 21, 25	Divide by 4, R = 1
17, 19, 23, 29	Prime numbers; differ by 1 from a multiple of 3
18	Evenly divided (R = 0) by 2, 3, and 9
20	Even number and multiple of 4 and 5. Number of fingers and toes
21	Evenly divided by 7 and 3; (R = 0) sum of digits is 3; a triangular number (ie, can be arranged like a bowling-pin triangle)
22	Multiple of 11; its two digits are the same.
24	Two dozen; divisible by 2, 3, 4, 6, 8, 12
25	A square number (5 rows of 5); one-fourth of 100
26	Even number; multiple of 13; half the number of weeks in a year, or cards in a deck.
27	3 x 3 x 3 and a multiple of 9
28	Four 7's; days in the shortest month of the year; a triangular number (arrange like bowling pins to make a triangle)

et cetera

jottings

Many times every day you run into things that can be divided into two similar halves: a dog, a person, a butterfly, a car, and many buildings, just to name a few examples. The quality they all have is called **symmetry.** Here are some symmetry activities that should give you some fun as you explore this idea in different ways. You and your teacher can decide on one that seems interesting and challenging to you. (They get harder as they go.)

Copycat game for two

Use the **Copycat** sheet with large squares to play this cooperative game. Make sure you have lots of colored pens or crayons.

🔍 *Note that this game can be played as solitaire by one student making both moves.*

RULES

• Pick two colors and color any two squares on the **left** side of the line.
• Your partner must use the same colors for the two squares on the **right** side that are in the **same rows** and the same **distances** from the line.
• Keep doing this until the paper is completely colored.

Congratulations, you've made an interesting symmetrical design!

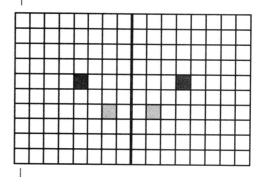

Mirror-Sculpture cooperative group game

RULES

• Join together in a group of four students — standing or sitting — and arrange yourselves in a *symmetrical* way.
• When you're ready, name a spot where someone, a "viewer," can stand to view your group "sculpture" so it looks symmetrical from there. That is, it has an imaginary center line.
• The viewer-person says "mirror" if you have succeeded, or says something like "adjust legs" or "move heads" if you don't have symmetry yet.

🔍 *Arms must be somewhere out from the body, and legs must not be straight down.*

🔍 *The arrangement on each side of that line should look like the arrangement on the other side.*

helpful | to do | calculate | write

COPYCAT

Nosymm game for two

This is a cooperative game where you and your partner are creating a design as you play. Use the **Copycat** Sheet.

RULES

- One partner agrees to be "Symm" and the other becomes "Nosymm."
- Symm starts with **two** colors and colors **four** squares (anywhere on either side of the center line) so they form a **symmetrical** pattern.
- Nosymm wrecks the symmetry of the pattern by coloring **two** squares **anywhere** on the sheet with **any** two colors.
- Symm must then make the pattern **symmetrical** again by coloring **two** squares. Then Nosymm wrecks the symmetry by coloring **two** squares.
- Continue until no more squares can be colored or until Symm can't fix the design.

Rotate game for two

Using the **Rotate** Sheet you can play the cooperative game "Rotate" with a partner.

RULES

- You color a square somewhere on the *left* of the center line.
- Your partner must color a square the same distance to the **right** of the solid center line and **below** the dashed line as yours was to the **left** of the solid line and **above** the dashed one (or if you were below the dashed one, your partner's would be above it).
- Continue until the paper is completely colored.

At any time during your play, put one finger on the center point and spin the paper halfway around it until **your** colors are on your partner's side and your **partner's** colors are on your side. What do you notice? _____

_____

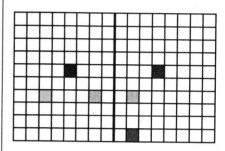

Note that this game can be played as solitaire by one student making both moves.

*Here's a good **variation**: Nosymm colors **three** squares with **three** colors after Symm starts with **four** squares as before. Then Symm continues with three.*

Note that this game can be played as solitaire by one student making both moves.

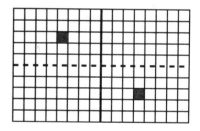

*Your design doesn't have "line-symmetry." It won't fold on top of itself. But it will spin halfway around to look like itself. This is called "**point-symmetry**."*

helpful

to do

calculate

write

ROTATE

Line Design game

Use the **Line Design** Sheet covered with dots. Play this cooperative game with a partner. You will need a **ruler**.

RULES

- Draw a straight line connecting **any** two dots on the paper. Your line can be on one side of the center line or can cross it.
- Your partner must draw another line that **balances** the first so that you have an image that is symmetrical around the center line.
- Continue taking turns for at least ten plays or until you have an interesting design. (Coloring the design in some way can make it even more interesting.)

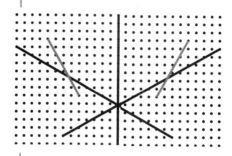

Note that this game can be played as solitaire by one student making both moves.

Line Design Nosymm game

RULES

- One partner agrees to be "Symm" and the other to be "Nosymm."
- Symm starts with **two** line segments anywhere on the paper, making a pattern that is **symmetrical** on each side of the line.
- Nosymm makes **two** line segments that **wreck** the symmetry.
- Symm must make two lines that **restore** the symmetry.
- Continue for ten turns each. Color all or part of the design you've made.

Note that this game can be played as solitaire by one student making both moves.

Rotate Lines game for two (Use after the Rotate and Line Design Games)

Use the **Line Design** Sheet. Play this cooperative game with a partner. This is challenging!

RULES

- Make a segment by connecting **any** two dots on the paper.
- Then your partner (or you) should make a **point-symmetric** image of that segment. That is, connect two dots with a segment on the **opposite side** of the center dot.
- Take turns making lines and point-symmetric images at least eight times. Check lines by spinning the paper halfway around — is the image in the same place as your partner's line?

Note that this game can be played as solitaire by one student making both moves.

*Another check to see if your image is in the correct place is that a ruler touching **your line's endpoint and the center large dot** will reach to touch the endpoint of the image. This must work for **both** endpoints of your line.*

 Experiment. On another piece of dot-paper make point-symmetric figures so that they form an interesting design.

 helpful to do calculate write

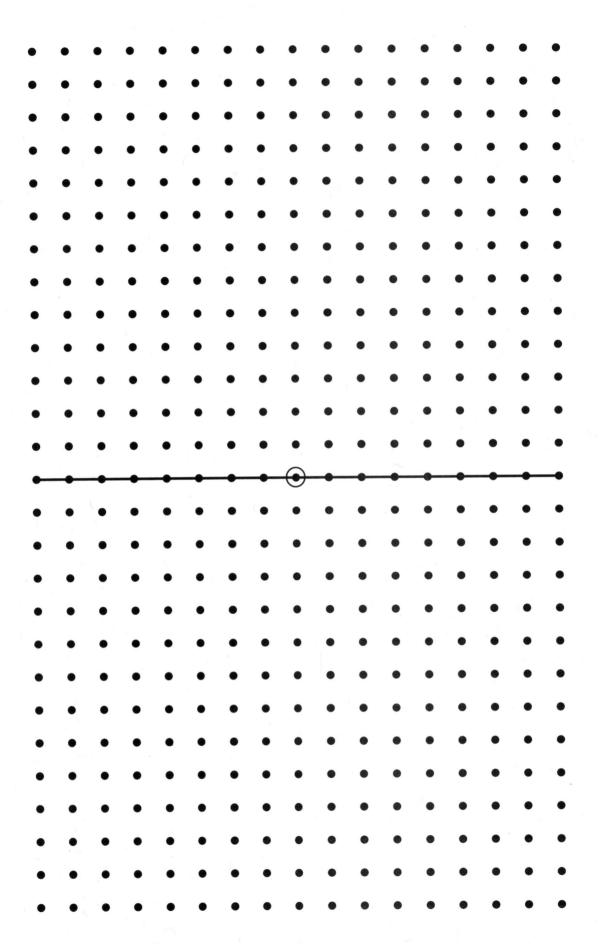

LINE DESIGN

Teacher Notes

Intelligences Emphasized: Visual-Spatial, Kinesthetic, Interpersonal

Math Skills, Concepts: Symmetry, spatial orientation

Materials: Copies of the "Copycat," "Rotate," and "Line Design" sheets, colored pens or crayons, rulers, instruction sheets for games

Level of Challenge: Easy to moderately challenging

Overview: These activities develop a physical, visual feeling for symmetry of both line and point types. Symmetry is one of the "big ideas" in geometry and in the comprehension of pattern. It is a quality that pervades both nature and human creations. Found in art and architecture, symmetry extends into the numerical, musical, intellectual, and metaphorical realms as well.

There are symmetrical *numbers* (like 34,543) and symmetrical *sentences* (like "Madam, I'm Adam."). Both are called *palindromes*. There are *rhythms* (like .. .. / .. .. / .. .. repeating) and *musical arrangements* (like a high instrument going *up* a progression of notes while a low instrument goes *down* a similar progression) that sound and feel symmetrical. There are symmetrical *concepts* like *reciprocity* and *balance*. Confucian philosophy speaks of symmetrical universal *energies* like yin and yang (symbolized by a symmetrical symbol), and the Christian ethic speaks of the symmetrical need to "Love your neighbor as yourself." The extensions below will give you a sense of how to explore symmetry further.

Activities 1 through 3 deal with this kind of two-way symmetry. It is the first type they encounter and it is called **bilateral, "mirror," or reflectional symmetry**. A characteristic of it is that one can place a mirror on a line across the figure so that the image in the mirror re-creates the other half. (Actual mirror pieces can be procured from a glass store, by the way, as a "donation to children at our school, tomorrow's leaders." Be sure to tape the sharp edges of the mirrors.) A figure on a flat plane can be folded along its line of symmetry to exactly lie over itself. One figure can have several such lines of symmetry, as in the case of a regular hexagon.

During or after the "Rotate" game of Activity 4 it is important to more clearly distinguish the second major type of symmetry — **point or rotational symmetry**. The design formed in Activity 4 is not line-symmetrical — no fold will cause it to lie on its other half. *But,* if it's rotated around its center half a turn, it will look exactly the same again. Some figures, like an equilateral triangle, need to be rotated only 1/3 of a turn to look the same again; others, like a left- or right-spinning swastika, need only 1/4 of a turn. It's important to show examples of these, even among letters of the alphabet (with its point-symmetrical N, S, and Z), and to have students make some of their own.

Pairs of students may want to peruse the activities and pick ones that fit their interest. There is no need to do them in order. A follow-up discussion of concepts learned and of broader ideas of symmetry would be instructive. Then students could be introduced to the extensions given at the end of these notes.

Introduction to Class: The **Overview** above offers versions of symmetry that address the different intelligences; these will help you to join with the students' emerging concepts of symmetry. Be sure to introduce and discuss symmetry in several different ways, drawing liberally on student-generated examples.

Because Activities 1 and 2 are good introductions to symmetry for less experienced students, an alternative approach would be to offer them first with only minor discussion, then follow with the symmetry discussion mentioned in the paragraph above. Activity 3 would then reinforce the concept. Activity 4 introduces **point symmetry**, which is at first very challenging until the student starts to think **rotationally.** Then activities 5 and 6 work with these types using lines.

The Rotate Lines game is very challenging and many will revert to line symmetry as they seek to draw lines. Demonstrate the principle mentioned there: a ruler running from an **end** of the **starter segment** through the **center point** will automatically touch an **end** of the **image segment** AND this must work for the **other end** of the starter as well. You can also check that segments and their images are equidistant from the point and also from the edges of the outside borders.

Extensions:
Alphabet. Make a vertical list of all of the letters of the alphabet. Next to each indicate whether it has "mirror" or "point" symmetry or "none." If it is *mirror symmetric,* like H, and there are two lines on which it could be folded, write "Mirror, 2." If it is *point symmetric* and it must be spun *halfway* around to look like itself again, like

MARKЯЯAM

MARK
WARK

Answers to Figure-Drawing

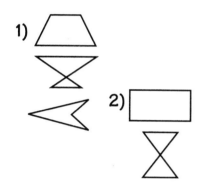

1)

2)

3) No figure possible

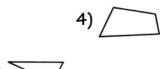

4)

5)

N, write "point, half." Are there any that must be spun only a *third* of the way to look like itself? [No.]

World Alphabets. Use Greek capital letters and do the same as above. [The delta is a "point, third" letter. A study of some of the other alphabets of the world and their symmetries (or lack thereof) would be fascinating here. Other symbols and designs can be studied also.]

Flip-Over Mirror-Writing. Make a mirror copy of your first name (and other words) by printing it in capitals then printing it backwards with letters reversed so that you truly have a mirror image.

Flip-Up Mirror-Writing. Make a mirror copy of your name by printing it right side up with a reflection of it upside down.

Figure-Drawing. Using four line segments, make figures that have 1) only one line of symmetry; 2) two lines; 3) three lines; 4) no lines, and 5) only point symmetry. (See the answers in the margin. They're examples of the correct responses.)

Treasure Hunt. Have a symmetry treasure hunt outside or in the school building or for homework in the city. The student must list (draw) various objects or designs encountered and indicate what kind of symmetry they have ("mirror, 2"; "point third," etc., as in the first extension).

Logo Hunt. Search through the yellow pages for logos (trademarks) of companies. Trace them and indicate what kind of symmetry they have.

Handshake Symmetry. What kind of symmetry does a handshake have? With a group of three try to form three hands in a handshake. Can you create a point-symmetric figure that would reproduce itself in one-third of a turn?

Mirror-Partner. Play MIRROR in pairs. Two students stand facing each other. One student leads, slowly moving arms, face, legs, or whole body. The other follows to make a mirror image. Then switch leadership. [Playing music during it is conducive to lowering self-consciousness.]

Nature Symmetry. Stimulate the naturalist intelligence with a study of nature's use of symmetry. Use field trips, collections, magazine picture collages, images from the microscopic world, etc., to explore nature's play with rotational and bilateral symmetry.

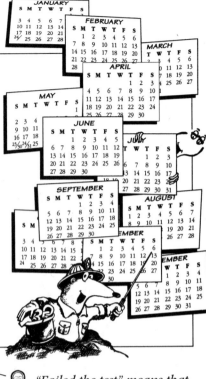

CALENDAR CONNECTIONS

For thousands of years people have thought that calendars are very important. A calendar tells when to expect the seasons, how many weeks have gone by, and when important celebrations are to be held. In fact, with the calendar in this activity you will be able to tell which day your birthday will be on in any year. And you'll be able to figure out the week-day anyone you know was born on!

Calendar Counting

"Thirty days has September,
April, June, and November;
Thirty-one have all the rest,
Except Feb, which failed the test."

 a. Use this well-known rhyme to compute the number of days in a year _____.

 b. In the space on the right, **show** how you calculated in **a**. **Label** your calculations clearly so that anyone could understand how you found your answer.

 c. Does *every* year have the number of days found in **a**? _____ To find out, check all the **Februarys** in your **4-Year Calendar**. Look up "leap year" in your dictionary or ask someone about it. Write your findings here:

_____

_____

 d. Choose any month on your calendar that has **five Tuesdays**. Write their dates (numbers) here: ___, ___, ___, ___, ___.
What pattern do you see? _____

_____

 e. Circle one of these years: 1998 1999 2001. Use the calendar of that year to tally how many of **each day** there are in that year:
Sundays ____ Mondays ____ Tuesdays ____ Wednesdays ____
Thursdays ____ Fridays ____ Saturdays ____

 f. Add up all of your tally numbers. Does the total agree with **a**? _____

 g. Only **one** of your days should have a different tally. Does it? _____ Why does this happen? _____

_____

"Failed the test" means that February has only 28 days!

*It's OK to use a calculator for this, but **write down** and **label** what you're calculating right here:*

*Use **shortcuts** in your counting.*

*If more than one day in **d** has a different tally, re-check.*

helpful   to do   calculate   write

4-Year Calendar

JANUARY
S	M	T	W	T	F	S
				1	2	3
4	5	6	7	8	9	10
11	12	13	14	15	16	17
18	19	20	21	22	23	24
25	26	27	28	29	30	31

FEBRUARY
S	M	T	W	T	F	S
1	2	3	4	5	6	7
8	9	10	11	12	13	14
15	16	17	18	19	20	21
22	23	24	25	26	27	28

MARCH
S	M	T	W	T	F	S
1	2	3	4	5	6	7
8	9	10	11	12	13	14
15	16	17	18	19	20	21
22	23	24	25	26	27	28
29	30	31				

APRIL
S	M	T	W	T	F	S
			1	2	3	4
5	6	7	8	9	10	11
12	13	14	15	16	17	18
19	20	21	22	23	24	25
26	27	28	29	30		

MAY
S	M	T	W	T	F	S
					1	2
3	4	5	6	7	8	9
10	11	12	13	14	15	16
17	18	19	20	21	22	23
24/31	25	26	27	28	29	30

JUNE
S	M	T	W	T	F	S
	1	2	3	4	5	6
7	8	9	10	11	12	13
14	15	16	17	18	19	20
21	22	23	24	25	26	27
28	29	30				

JULY
S	M	T	W	T	F	S
			1	2	3	4
5	6	7	8	9	10	11
12	13	14	15	16	17	18
19	20	21	22	23	24	25
26	27	28	29	30	31	

AUGUST
S	M	T	W	T	F	S
						1
2	3	4	5	6	7	8
9	10	11	12	13	14	15
16	17	18	19	20	21	22
23/30	24/31	25	26	27	28	29

SEPTEMBER
S	M	T	W	T	F	S
		1	2	3	4	5
6	7	8	9	10	11	12
13	14	15	16	17	18	19
20	21	22	23	24	25	26
27	28	29	30			

OCTOBER
S	M	T	W	T	F	S
				1	2	3
4	5	6	7	8	9	10
11	12	13	14	15	16	17
18	19	20	21	22	23	24
25	26	27	28	29	30	31

NOVEMBER
S	M	T	W	T	F	S
1	2	3	4	5	6	7
8	9	10	11	12	13	14
15	16	17	18	19	20	21
22	23	24	25	26	27	28
29	30					

DECEMBER
S	M	T	W	T	F	S
		1	2	3	4	5
6	7	8	9	10	11	12
13	14	15	16	17	18	19
20	21	22	23	24	25	26
27	28	29	30	31		

JANUARY
S	M	T	W	T	F	S
					1	2
3	4	5	6	7	8	9
10	11	12	13	14	15	16
17	18	19	20	21	22	23
24/31	25	26	27	28	29	30

FEBRUARY
S	M	T	W	T	F	S
	1	2	3	4	5	6
7	8	9	10	11	12	13
14	15	16	17	18	19	20
21	22	23	24	25	26	27
28						

MARCH
S	M	T	W	T	F	S
	1	2	3	4	5	6
7	8	9	10	11	12	13
14	15	16	17	18	19	20
21	22	23	24	25	26	27
28	28	30	31			

APRIL
S	M	T	W	T	F	S
				1	2	3
4	5	6	7	8	9	10
11	12	13	14	15	16	17
18	19	20	21	22	23	24
25	26	27	28	29	30	

MAY
S	M	T	W	T	F	S
						1
2	3	4	5	6	7	8
9	10	11	12	13	14	15
16	17	18	19	20	21	22
23/30	24/31	25	26	27	28	29

JUNE
S	M	T	W	T	F	S
		1	2	3	4	5
6	7	8	9	10	11	12
13	14	15	16	17	18	19
20	21	22	23	24	25	26
27	28	29	30			

JULY
S	M	T	W	T	F	S
				1	2	3
4	5	6	7	8	9	10
11	12	13	14	15	16	17
18	19	20	21	22	23	24
25	26	27	28	29	30	31

AUGUST
S	M	T	W	T	F	S
1	2	3	4	5	6	7
8	9	10	11	12	13	14
15	16	17	18	19	20	21
22	23	24	25	26	27	28
29	30	31				

SEPTEMBER
S	M	T	W	T	F	S
			1	2	3	4
5	6	7	8	9	10	11
12	13	14	15	16	17	18
19	20	21	22	23	24	25
26	27	28	29	30		

OCTOBER
S	M	T	W	T	F	S
					1	2
3	4	5	6	7	8	9
10	11	12	13	14	15	16
17	18	19	20	21	22	23
24/31	25	26	27	28	29	30

NOVEMBER
S	M	T	W	T	F	S
	1	2	3	4	5	6
7	8	9	10	11	12	13
14	15	16	17	18	19	20
21	22	23	24	25	26	27
28	28	30				

DECEMBER
S	M	T	W	T	F	S
			1	2	3	4
5	6	7	8	9	10	11
12	13	14	15	16	17	18
19	20	21	22	23	24	25
26	27	28	29	30	31	

4-Year Calendar

JANUARY
S	M	T	W	T	F	S
						1
2	3	4	5	6	7	8
9	10	11	12	13	14	15
16	17	18	19	20	21	22
23/30	24/31	25	26	27	28	29

FEBRUARY
S	M	T	W	T	F	S
		1	2	3	4	5
6	7	8	9	10	11	12
13	14	15	16	17	18	19
20	21	22	23	24	25	26
27	28	29				

MARCH
S	M	T	W	T	F	S
		1	2	3	4	
5	6	7	8	9	10	11
12	13	14	15	16	17	18
19	20	21	22	23	24	25
26	27	28	29	30	31	

APRIL
S	M	T	W	T	F	S
						1
2	3	4	5	6	7	8
9	10	11	12	13	14	15
16	17	18	19	20	21	22
23/30	24	25	26	27	28	29

MAY
S	M	T	W	T	F	S
	1	2	3	4	5	6
7	8	9	10	11	12	13
14	15	16	17	18	19	20
21	22	23	24	25	26	27
28	29	30	31			

JUNE
S	M	T	W	T	F	S
				1	2	3
4	5	6	7	8	9	10
11	12	13	14	15	16	17
18	19	20	21	22	23	24
25	26	27	28	29	30	

JULY
S	M	T	W	T	F	S
						1
2	3	4	5	6	7	8
9	10	11	12	13	14	15
16	17	18	19	20	21	22
23/30	24/31	25	26	27	28	29

AUGUST
S	M	T	W	T	F	S
		1	2	3	4	5
6	7	8	9	10	11	12
13	14	15	16	17	18	19
20	21	22	23	24	25	26
27	28	29	30	31		

SEPTEMBER
S	M	T	W	T	F	S
					1	2
3	4	5	6	7	8	9
10	11	12	13	14	15	16
17	18	19	20	21	22	23
24	25	26	27	28	29	30

OCTOBER
S	M	T	W	T	F	S
1	2	3	4	5	6	7
8	9	10	11	12	13	14
15	16	17	18	19	20	21
22	23	24	25	26	27	28
29	30	31				

NOVEMBER
S	M	T	W	T	F	S
		1	2	3	4	
5	6	7	8	9	10	11
12	13	14	15	16	17	18
19	20	21	22	23	24	25
26	27	28	29	30		

DECEMBER
S	M	T	W	T	F	S
					1	2
3	4	5	6	7	8	9
10	11	12	13	14	15	16
17	18	19	20	21	22	23
24/31	25	26	27	28	29	30

JANUARY
S	M	T	W	T	F	S
	1	2	3	4	5	6
7	8	9	10	11	12	13
14	15	16	17	18	19	20
21	22	23	24	25	26	27
28	29	30	31			

FEBRUARY
S	M	T	W	T	F	S
				1	2	3
4	5	6	7	8	9	10
11	12	13	14	15	16	17
18	19	20	21	22	23	24
25	26	27	28			

MARCH
S	M	T	W	T	F	S
				1	2	3
4	5	6	7	8	9	10
11	12	13	14	15	16	17
18	19	20	21	22	23	24
25	26	27	28	29	30	31

APRIL
S	M	T	W	T	F	S
1	2	3	4	5	6	7
8	9	10	11	12	13	14
15	16	17	18	19	20	21
22	23	24	25	26	27	28
29	30					

MAY
S	M	T	W	T	F	S
		1	2	3	4	5
6	7	8	9	10	11	12
13	14	15	16	17	18	19
20	21	22	23	24	25	26
27	28	29	30	31		

JUNE
S	M	T	W	T	F	S
					1	2
3	4	5	6	7	8	9
10	11	12	13	14	15	16
17	18	19	20	21	22	23
24	25	26	27	28	29	30

JULY
S	M	T	W	T	F	S
1	2	3	4	5	6	7
8	9	10	11	12	13	14
15	16	17	18	19	20	21
22	23	24	25	26	27	28
29	30	31				

AUGUST
S	M	T	W	T	F	S
			1	2	3	4
5	6	7	8	9	10	11
12	13	14	15	16	17	18
19	20	21	22	23	24	25
26	27	28	29	30	31	

SEPTEMBER
S	M	T	W	T	F	S
						1
2	3	4	5	6	7	8
9	10	11	12	13	14	15
16	17	18	19	20	21	22
23/30	24	25	26	27	28	29

OCTOBER
S	M	T	W	T	F	S
	1	2	3	4	5	6
7	8	9	10	11	12	13
14	15	16	17	18	19	20
21	22	23	24	25	26	27
28	29	30	31			

NOVEMBER
S	M	T	W	T	F	S
				1	2	3
4	5	6	7	8	9	10
11	12	13	14	15	16	17
18	19	20	21	22	23	24
25	26	27	28	29	30	

DECEMBER
S	M	T	W	T	F	S
						1
2	3	4	5	6	7	8
9	10	11	12	13	14	15
16	17	18	19	20	21	22
23/30	24/31	25	26	27	28	29

Calendar Patterns

 a. On *any* calendar month pick a group of numbers that makes a *square*. **Lightly** draw a square on them with your pencil. See the examples of different squares possible on the right.

 b. Add all the numbers along each **diagonal** of the square and on the middle column, as shown at the right. Write the three totals here: ___, ___, ___ Pick at least **four more squares** in that same month, and add their diagonals and middle rows. Write your totals.

 c. Describe what was true of **all** the squares. Write your results in a complete sentence that starts with these words: In a square of calendar numbers, adding the diagonal numbers and middle row numbers_____

_____

 d. Why does the pattern in **c** work? _____

_____

_____

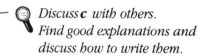 *Discuss **c** with others. Find good explanations and discuss how to write them.*

 e. In a month of your choice, pick five two-by two squares. **Multiply** each diagonal of each square and write the answers here:

square 1 ___, ___   square 4 ___, ___
square 2 ___, ___   square 5 ___, ___
square 3 ___, ___

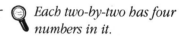 *Each two-by-two has four numbers in it.*

 f. What is the same relationship that all five pairs of numbers have? _____

_____

 g. Look for other patterns in the numbers on a calendar page. Show them and describe them on another paper.

Christmas Patterns

a. Find **Christmas** (December 25) on each year of the **4-Year Calendar**. List what days it falls on in those four years.

_____, _____, _____, _____

helpful   to do   calculate   write

b. Is there a pattern? _____ If 2000 were **not** a leap year (with one extra day in February) what day would Christmas be on? _____ Describe the pattern the four Christmas days would make **then**: _____

The extra day in February moves every date one day later for the rest of the year.

c. Try this same investigation for the days of your **birthday** during the four years.

Chip-Calendar

a. Make a moving calendar page by numbering 31 "chips" (tokens, paper squares, poker chips, or cubes) with big numbers 1-30. Write on stickers on the chips or right on the papers to make the numbers show up. Use the **Chip Calendar** page and arrange your chips as discussed in **b**.

Chip Calendar

Sunday	Monday	Tuesday	Wednesday	Thursday	Friday	Saturday

b. Arrange 31, 30, 29, or 28 chips in the calendar to show **this month**. Put a strip of paper above the page with this month's name. Try to rearrange the chips to make **next month** (or any other month you choose) without looking on a calendar.

(Challenging) Thanksgiving Puzzler
The United States government has ruled that Thanksgiving must be on the **fourth Thursday in each November**.

a. Thanksgiving dates are the 23d in 1995, the 28th in 1996 (a leap year), and the 27th in 1997. Use the **4-Year Calendar** and write the dates **Thanksgiving** occurs on in November from 1998 to 2001: _____, _____, _____, _____.

b. What is the **earliest** date in November that Thanksgiving can occur? _____
What is the **latest** date that it can occur? _____

*If you want to explore and be successful with questions **b-e**, use a **Chip-Calendar** (from the previous activity) for November.*

c. How does the Thanksgiving date change in the leap years 1996 and 2000? _____

d. How does the Thanksgiving date change from year to year?
_____
_____

*Try different possibilities with the **Chip-Calendar** until you see the relationships. Think of how its date depends on what day November 1 is.*

e. What day was Thanksgiving on in 1994? _____ Predict what day it will be on in the year 2002. _____

helpful   to do   calculate   write

Chip Calendar

Sunday	Monday	Tuesday	Wednesday	Thursday	Friday	Saturday

Predict 2004 from the 4-Year Calendar

 a. On what *days* will February of 2004 start and end? _____,
_____.

 b. On what day will the year end? _____

 c. Which day(s) of the week will occur most often in
2004? _____

Remember, 2004 is a leap year.

What Day of the Week?

Here's a challenging trick for finding on what **day of the week**
any date in the past or future occurs. Youll use all three tables at
the right.

Example: Spring Equinox, March 21, 1999.

Simply find numbers for each of the **steps 1)** through **6)** and
add them up, then **divide** the answer by 7.

STEPS

**1) Divide 12 into the last two digits of the year.
Add the answer and remainder.**

2) How many 4s fit in the remainder?

3) Add the MONTH CODE.

4) Add the day of the month.

**5) Is the year in the 2000s or
a leap year? If not, write + 0.
If yes, go to the "2000s OR LEAP YEAR?" box.**

6) Find the total of your column of numbers.

**7) Divide the total by 7. The remainder means the day. (In
this case the remainder 1 means March 21, 1999 is Sunday.)**

$$12\overline{)99} \quad 8\,R\,3$$

	8
	+ 3
There are **0** 4s in 3.	+ 0
For March it's **4**	+ 4
Here it's **21**	+ 21
	+0

$$7\overline{)36} \quad 5\,R\,1$$

36

MONTH CODES	
January	1
February	4
March	4
April	0
May	2
June	5
July	0
August	3
September	6
October	1
November	4
December	6

DAY NUMBERS (USE REMAINDER)	
Sunday	**1**
Monday	2
Tuesday	3
Wednesday	4
Thursday	5
Friday	6
Saturday	0

 a. Follow **steps 1)** through **7)** above for a date on your 4-Year
Calendar. Make sure the day you get is right.

 b. Use the **STEPS** and find the day of the week you were born
on. _____

 c. Birth-day Return: Figure out in which year or years your
birthday will occur on the same day of the week you were
born on. That is, if you're born on Monday, when will your
birthday be on Monday again?

2000s OR LEAP YEAR?

In step 5),

- For **any** 2000s leap year
  before Mar. 1, write -**2**
- For **any** 2000s leap year
  Mar. 1 or after, write -**1**
- For **any other** 2000s years,
  write -**1**
- For **any** 1900s leap year
  before Mar. 1, write -**1**
- For **any** <u>1900s leap year</u>
  Mar. 1 or after, write +**0**

helpful   to do   calculate   write

d. Day Finder Cooperative Game. Take turns with a partner being the "Day Finder." The partner asks the Day Finder a **date** from the **4-Year Calendar**, without saying the day of the week. The Day Finder computes the day and the partner tells whether it's correct. If so, the partner becomes Day Finder. If it's not, the Day Finder has one more chance to get it right before switching.

e. Class Birthday Graph and Pictures. You, or each student in class, can compute and list on the board the day of the week each student was born on. Make a **bar graph** that shows the student total for Sunday, Monday, etc.

All the days are named from **ancient myths**. You, your group, or the class, can draw pictures for each of the days based on their mythical meanings in the chart below. The "most popular" day on the graph gets the largest picture!

The "most popular" item on a bar graph is called its "mode."

To get picture ideas, you can do research and find pictures of the gods and goddesses the days are named after. These come from Norse, Roman, and Anglo-Saxon mythology.

Day	Named After	More Information
Sunday	The sun	There was a holiday (holy day) called "Sun's Day" in the Roman calendar and later many people made it the name of the first day.
Monday	The moon	Monday means "moonday". The Anglo-Saxons felt that the second day of the week belonged to the goddess of the moon.
Tuesday	Tiu (also called Tyr)	Tuesday is Tiu's day. Tiu is the Norse god of war, similar to Mars of the Romans.
Wednesday	Wodin (also spelled Odin)	Wednesday is Wodin's day. Wodin was king of the Norse gods, who rode an eight-footed horse and had a powerful ring. He was also god of wisdom, poetry, and magic.
Thursday	Thor	Thursday is Thor's day. Thor was the Norse god of thunder. He also loved iron things.
Friday	Frija or Fria	Friday is Fria's day. Fria is the Norse goddess of love similar to Venus of the Romans.
Saturday	Saturn	Saturday is Saturn's day. Saturn was the Roman god of the sky and of farming.

helpful to do calculate write

Teacher Notes

Intelligences Emphasized: Spatial, Bodily-Kinesthetic, Linguistic

Math Skills, Concepts: Systematic counting, number patterns, mental calculation, simple division and use of remainders, logical sequences of operations

Materials: Ruler, colored pens, calculator

Level of Challenge: Moderately easy to moderately challenging

Overview: This activity looks at many aspects of calendars, with the overall goal of creating greater familiarity with the structure of calendars and the flow of time. In the process, students learn to search for patterns among numbers and dates, hypothesizing about their causes and predicting from them.

Young people are fascinated with their birthdays. Several activities involve birthdays and these could be used as a "hook" to entice more calendar exploration afterwards.

Introduction to class: Some possible lead-ins to this activity are these:

1) Allow students to peruse the 4-Year Calendar pages and discuss what they notice.

2) Ask students general questions about calendars as you hold a current one in your hand. (Questions may arise that you don't know the answers to, which could incite a research assignment before starting the Activity.) What are they for? How old is our modern calendar structure? Does everyone in the world use this calendar? What have other calendars been like? How does the calendar change from year to year? How are months different? What do calendars have to do with astronomy?

3) Have students work on the first page of the Activity then carry on the discussion in 2).

4) Start with singing the musical scale "do, re, mi, fa, so, la, ti" then "do, re" again. How do the second "re" and the first "re" relate? (Same note but higher.) Then segue to the week, singing the days as the scale, and pointing to the days on a calendar page, "Sunday, Monday, ... , Saturday" then "Sunday, Monday, ..." again. How does the second Monday relate to the first Monday? Same day, later date. Continue by showing how the whole year is just more "octaves" of a week. Then start the Activity.

jottings

Answers

Calendar Counting

a. 365 days. **c.** Leap year puts in one more day on February 29, making those years have 366 days. **d.** The Tuesday dates increase by 7 each time.

e-g. 52 weeks **x** 7 days makes only 364 days. If a year had only 364 days, and started on a Monday, it would end on a Sunday, the day before, so that exactly 52 weeks would have occurred. But there's one more day to make 365, causing the year to end on the *same* day it began, and that's the day that occurs 53 times. If a *leap year* were tallied, there would be *two* days that occur 53 times.

Calendar Patterns

a-d. Diagonals and the middle row of any square will create the same totals, though the totals will differ from square to square. (Some sharp investigators may wish to study how totals change from square to square on the calendar page.) This works because whatever the numbers in the middle row are, each diagonal has numbers that are one more, equal to, and one less than the middle row numbers. **e-f.** For two-by-two number squares, diagonal products will differ by 7.

 Christmas Patterns, a-c. Each year Christmas occurs one day later (because every normal year is exactly 52 weeks and one extra day) or two days later in a leap year. This will be the same for students' birthdays.

 Thanksgiving Puzzler.

a. Thanksgiving dates:

Year:	1995	1996	1997	1998	1999	2000	2001	2002
Nov.	30th	28th	27th	26th	25th	30th	29th	28th

b. Earliest possible date: 22nd   Latest possible date: 28th

The best way to study this is with a **chip-calendar** with November arranged on it starting on Sunday (see the light-bulb "**Chip-Calendar**" activity). That puts Thanksgiving on the 26th. Move the 30 chips so that November starts on Monday, then Tuesday, then Wednesday, then Thursday; Thanksgiving moves to the 25th, the 24th, the 23rd, and the 22nd.

When November starts on a **Friday**, a shift happens. The fourth Thursday suddenly becomes the 28th.

Then Saturday, Sunday, and Monday, start-offs for November move Thanksgiving back to the 27th, 26th, and 25th. That would complete the story, except for leap years, which cause Thanksgiving's date to move back *two* days from the previous year's. In 1996 this happened, and even in 2000, when Thanksgiving should have been on the 24th, it did its leap to the 23rd one year early.

Predict 2004

a. February will start on a Sunday, three days later than it did in 2001, because *every* date of every year starts a day later than it did the year before. But February will **end two days later** than it did the year before because it has 29 days. Thus, it will end on a Sunday also, which is four weeks and one day later. **b.** Most years end on the same day they started. 2004 will end on Friday, one day later than the Thursday it started on (as do all leap years). **c.** Exercises **d-f** on the first page show that the day on which a year starts and ends is the extra day each year has. The leap year 2004 has two extra days, Thursday and Friday.

What Day of the Week? This method, while a bit challenging, simplifies when done a few times. Adjustments are made for leap years and dates in the 2000s (making the method Y2K compatible!) It involves finding six numbers for any date, adding those, then dividing by 7. The remainder tells what day of the week the date corresponds to. Subactivities **a - d** build this practice, with **c** optional. **c. Birth-day Return.** Kids get pretty fascinated with the idea of having a special birthday which is on the same day that they were born. One girl in my class, turning eleven, born on Monday, found out that her upcoming birthday would be Monday, that Monday was named after the moon, and that her upcoming birthday was on a full moon! She excitedly proclaimed that the moon would be the central theme of her party.

Students can make **charts** with three entries for each year: the year (starting from birth year), their age (starting from age 0), and the day of the week of the birthday each year. For each leap year the day of the week advances **two** days (if the birthdate is after February 29) or waits until the following year to advance two days if before February 29. It normally advances only one day in regular years. They will discover that birth-day returns

For this formula, students need to know how to find whether a year is a leap year. It's easy: the last two digits are a multiple of 4.

happen at ages 5 or 6, 11 or 12, 17 or 18, and beyond in increments of 6 years. **d. Class Birthday Graph and Pictures** opens the door to a study of ancient myths, art, and bar graphing.

Extensions

(Intercultural): [Instructions for the student] Research a few calendars of other cultures (Sumerian, Egyptian, Inca, Aztec, Mayan, Hebrew, Islamic, Old Roman, Chinese and East Indian are some possibilities) and prepare a presentation on their structures for the class. Be sure to use diagrams or charts to make it clearer. Also research the Gregorian Calendar (ours) and how it came to be.

(Linguistic): [Instructions for the student] Team up with others and make up a new calendar structure. Give it an interesting name. You can even write a fictional story about where it existed or how it was found.

For instance, it could be based on a different number of months, or weeks than ours. See if you can make one that seems to make more sense. The idea of seven days in a week probably came from the ancient Chaldeans — at any rate, it's very ancient. Try a different number of days in a week, and, for that matter a different number of hours in a day (meaning that your "hour" is longer or shorter than regular hours).

NOTE: Some things can't be changed. These are fixed by the sun-earth relationship. For instance, there are always 365 1/4 days (spins of the earth) for one trip around the sun (revolution) lasting one year. Winter and summer come only once a year. You may choose, as many cultures have, to base your calendar on the moon, with 12.3 cycles of 29 3/4 days between new moon and full moon each year. The "moonth" or month comes from this fact.

(Musical): [To be used by the teacher in conjunction with any exercises in the Activity] The musical scale has seven notes, just as the week has seven days. If Sunday is **do** then the next week starts with **do** again. A keyboard that has five scales on it can play the days of any month, starting on the note of the lowest scale that goes with the starting day of the month. Furthermore, any date of the month becomes a note. The **Calendar Patterns** activities can be *heard* as well as *seen* this way.

For instance, sounding notes corresponding to dates for a row or diagonal of any of the squares used in **Calendar Patterns** can bring new insights, especially to the musically intelligent. Notes of rows or diagonals can be sounded either simultaneously or sequentially for different "takes" on their sums or patterns.

FAIR AND SQUARE

The numbers 1 through 9 aren't just a line of cold symbols — they're part of a tight family. They get along so well with each other that they can weave together into very surprising patterns. They do it especially well in a **Magic Square.** In this activity you will learn to recognize and communicate patterns in these squares.

Magic Square

Here's how to weave one. In the diagram at the right the nine digits are nicely arranged in stairsteps.

 a. Let **1** jump into the gray square by leaping **over** the **5** and landing in the box below it. Then, let *each* outside number jump over the **5** into a gray box.

 After a number leaps, write it in the new box and cross it out in the old one.

The digits 1 through 9 are now arranged in tic-tac-toe form in the gray square. So what's the big deal about that? To find out, try these things:

 b. Color all the **odd**s in the gray square *yellow.* Look for a pattern. What is it? _____

c. Add up the **first row** of numbers. Write the total at the end of the row.

 d. Add the first **column** (4, 3, and 8) and write the total at the end of the column.

 e. Total each row and column the same way. Also total each **diagonal** and write these.

 f. What's true about all of the totals? _____

Ancient peoples of India and China thought of this arrangement as special, lucky, and magical for thousands of years.

helpful   to do   calculate   write

The Code Square

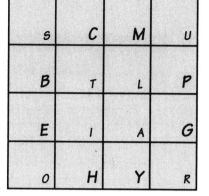

Making the Code Square (use a pencil)

The numbers 1 to 16 weave a 4-by-4 Magic Square also. It's as simple as 1, 2, 3 to follow their weaving. Here's how:

 a. In the square at the right, touch boxes **C, M, B, P, E, G, H, and Y** in order with your pencil or finger. Then do it again faster. Do you feel or see a pattern those boxes make? _____

 b. Start again at **S** and move **left to right** along each row, **touching** each square with your pencil while saying, "1, 2, 3, …" up to "16." But when you touch each of the boxes **C, M, B, P, E, G**, **H, and Y**, <u>write</u> the number you're saying.

As you touch the other boxes only <u>say</u> the numbers but <u>don't write</u> them.

Afterwards, check yourself: Did box G get a 12?

 c. Count from the **last** box, R, to the **first**, S, going along each row from **right** to **left**. Touch each box as you say 1 to 16 again. Each time you touch an empty box, write the number you're saying.

So, 1 goes in box R, 4 in O, 6 in A, 7 in I, until 16 is in S.)

 To see if you did it right, try adding each **row**, **column and diagonal**. What do you find? _____
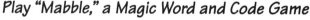

If all are not the same total, read the directions carefully and try making the square again.

Play "Mabble," a Magic Word and Code Game

Mabble will use the Magic "Code Square" you just made. It will challenge your spelling skill, your ability to see patterns, and your mental addition speed.

RULES

For each play:

- Find **four boxes** in the Code Square that occur in some **orderly arrangement** and that have a **sum of 34.**

An "orderly arrangement" looks like a pattern in the square. The four numbers make a nice shape or line.

- Say their four letters as a "**code**" so the other player(s) can check that your code makes a 34-sum.

- Work with your code and get a score for it. **Be clever** to get the most points you can by shrewdly using the **Letter Tricks and Scores Chart** on the next page!

helpful · to do · calculate · write

LETTER TRICKS AND SCORES CHART

LETTER TRICKS	SCORE
The code letters can be arranged to **spell a word.** Examples: SCUM, STAR	8 points
The code letters can be arranged to **rhyme** with a real word. Example: LAYM rhymes with lame.	5 points
Three letters of the code can be arranged to make a word. Example: UPGR has RUG.	4 points
You can make a word from the code by adding **one extra letter.** Example: SCYR + A makes SCARY.	4 points
Your four letters **don't make anything.**	2 points
You have **two useless codes** that scored only **2 points**. You can make **any sized word** with some letters from each of them. Example: CBGY and SBEO are 2-pointers. Together they can make BEGS. It gets 4 x 1/2, or 2 **bonus** points.	Bonus: 1/2 point per letter of new word.

You can't use any of the Examples given here to get points in your game!

- **Winning:** The highest score after 20 minutes of play wins, **or** the first person to get 50 points wins, **whatever comes first.**

The Rune Square and Games

You will be creating "runes" in this Activity. A rune is an ancient mysterious symbol that has secret, powerful meanings. The runes you make will resemble the original ones used thousands of years ago in Germany. They also resemble "hieroglyphs" used in Egypt and "logos," used by businesses (look up these words).

For the Rune Game you will use **another** Magic Square, shown on the next page, that has, like runes, been well-known for many centuries. Call it the **"Rune Square"**. It contains some 34s that the other one doesn't. On the next page are the rules for how you can play a game with a partner, or play solitaire.

RUNE RULES

- There is a strange **figure** in each box along with the **number**. Find any four nicely arranged numbers that add to 34.

- Make **a rune** that represents them by joining their **four figures** in an interesting way.

- To make the rune, each figure can be drawn **bigger or smaller**.

- Each figure **cannot be turned** to be used. It must be pointed the **same direction** in the rune as it is in its box.

- Each figure that makes a rune must be **connected to,** or **inside of,** another figure.

THE GAME RULES

- Partners take turns being **"Runemaker"**. For a play, the Runemaker finds four numbers in an arrangement of boxes that make a total of 34.

- The Runemaker creates a rune, following the RUNE RULES.

- The partner must look at this rune, and, on the first guess, identify what four numbers the Runemaker had in mind.

- If the partner can't say the four correctly, the Runemaker gets a point.

- If the partner can prove that the Runemaker broke the RUNE RULES, the partner gets a point instead.

- After eight turns each, the highest points win.

7 ⊃	12 ○	1 ⊏	14 ∪
2 ╱	13 ‿	8 ∧	11 ∿
16 ◎	3 ⊂	10 ‿	5 —
9 │	6 ╲	15 •	4 ◗

 For example, looking at this rune and the Rune-Square, can you tell that it says that 14 + 11 + 1 + 8 = 34?

A *Group-Runes Game* (For three or more)

RULES

All partners agree on an orderly arrangement of numbers that make 34. All partners privately make their own runes with the rune-pieces from those four boxes. All partners display their runes to the rest of the group. Then make a single "group-rune" that uses the best ideas from all. After doing this for several runes, the game group makes a **bulletin-board** display of their runes.

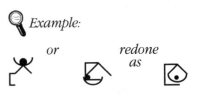

Example:

or   *redone as*

A Box of Patterns

Look at the page marked **A Box of Patterns.** On it is a 4-by-4 grid with sixteen of the same symbols in each box.

The 4-by-4 grid represents a magic square.

 a. If you played the games, go over your list of **letter-codes** from the "**Code Square**" or your **runes** from the "**Rune Square**" to recall the orderly patterns of numbers you found that make a sum of **34**.

 b. Color the **A Box of Patterns** for as many 34-arrangements of four boxes as you can. Look for arrangements that make 34 in either the Code Square or Rune Square. Always keep this rule:

*You need a pen assortment with **many** colors in it.*

GRID COLORING RULE

- Color the **same symbol** in each set of four chosen boxes the **same color**.

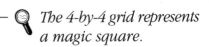

Example: *For the code CMTL color the **6-pointed star** shape inside boxes C, M, T, and L **pink**.*

 You can choose any symbol to color in your four boxes as long as **all four** of them are uncolored. Check that all are not colored **before** you start to color a group of four.

helpful   to do   calculate   write

A BOX OF PATTERNS

A Design Surprise

(You will need a ruler and colored pens)

At the right is the small three-by-three Magic Square you made first. You can make a neat design with it. **Just use each row as a recipe for connecting dots!** On the right are nine numbered dots. Here's how to connect them step by step:

4	9	2
3	5	7
8	1	6

3 x 3 Square

1   2   3
●   ●   ●

4   5   6
●   ●   ●

7   8   9
●   ●   ●

 a. Use the 3-by-3 square as your guide. Think of its first row (4, 9, 2) as saying, "In the dot-square at the right, join dots 4 to 9 to 2 with **straight** lines. Join them in order.

Use a ruler!

 b. The 3, 5, 7 row says, "Join 3 to 5 to 7." Go ahead.

 c. Join 8 to 1 to 6.

 d. Make lines **around** your design: Join 1 to 3 to 9 to 7 to 1 in order.

Your lines form a design that may be hard to appreciate until you color it.

 e. Color the shapes in the design with at least **three different colors** and you'll see how interesting the design is.

 f. Now find the **4-by-4 Code Square**. Use the square of dots at the right. Read that square's first row as if it says "Connect 16 to 2 to 3 to 13." Then connect 5 to 11, and so on. Finish by joining 16 to 13 to 1 to 4 to 16 to **frame** the design. Color this complex design with **at least three different colors**.

Code Square

1   2   3   4
●   ●   ●   ●

5   6   7   8
●   ●   ●   ●

9   10   11   12
●   ●   ●   ●

13   14   15   16
●   ●   ●   ●

 g. Do the same with the **Rune-Square** and the third set of dots below. Use the same method; color the design it makes.

 Make your own.
Make 16 dots in a square shape and make up **your own design** on them with a ruler. Color it.

Rune Square

1   2   3   4
●   ●   ●   ●

5   6   7   8
●   ●   ●   ●

9   10   11   12
●   ●   ●   ●

13   14   15   16
●   ●   ●   ●

helpful   to do   calculate   write

Musical designs. Label a selection of 16 notes on the piano, xylophone, or synthesizer with little stickers numbered 1 through 16. In one of the 4-by-4 squares select an arrangement of numbers that total 34. Play the four notes that go with these numbers. Do this for several such sets of four notes. Listen for relationships among their sounds.

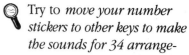 Try to *move your number stickers to other keys to make the sounds for 34 arrangements even more interesting.*

 Make a Zillion Magic Squares. Think of a small number. Multiply **each** number of any of your Magic Squares by that chosen number and put the answers in the boxes of a new Magic Square. Is this square still magic? Also try picking a number and **adding** it to each number in a Magic Square. Is the result a Magic Square? Try interchanging two rows of a Magic Square, like top and bottom. Is the result a Magic Square? What about columns? Try other experiments.

Numbers you use in these can even be fractions or decimals!

 A 5-by-5 Magic Square.

 a. Examine again on the first page how the 3-by-3 Magic Square was made. Notice the similar way that the numbers could have started out in this 5-by-5 square (as shown in gray). Figure out how they could have leaped into the square from the gray boxes.

b. Explore the sums in the Magic Square and look for other patterns that make the total.

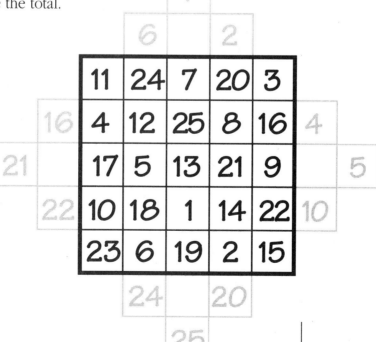

helpful | to do | calculate | write

Teacher Notes

Intelligences emphasized: Visual-Spatial, Linguistic, Interpersonal, Kinesthetic, Musical, Naturalist (sorting and classifying patterns)

Math skills, concepts: Mental addition, symmetry, visual patterns

Materials: Colored pens (many colors!), ruler

Level of Challenge: Moderately easy to moderately challenging

Overview: Magic squares have been studied mainly from the seventeenth century on, but they have been known in China as early as 2200 BC. A 4-by-4 square was first discovered in India in 1100 AD.

These activities are designed to reinforce mental calculation as children total various arrangements that make 34. They are also designed to appeal to as many as seven of the eight intelligences. In the language of learning styles (see Chapter 6), the dolphins and foxes will be especially pleased! Because of the variety here, you may wish to offer students or groups a choice of whether number groups are to be converted to words, to musical sounds, to visual glyphs, to colors, or to designs to interest them in the age-old number patterns of the Magic Square.

Answers and Tips

Making the Code Square and Playing "Mabble"

There are numerous symmetrical configurations in the Magic Square that total 34. For the Code Square I'll give you examples of the arrangements that are harder to find than rows, columns, and diagonals. Unless noted, these will work on the second, or Rune Square also. When I give you one code, realize that, unless noted, **each code given goes with a family of three other identical ones** in similar symmetrical positions in the square.

Here are sample codes for the orderly arrangements which, along with rows, columns and diagonals make 50 different codes in all:

Relatives of	Comments
SCBT	Note that CMTL and its relatives *won't* work here but *will* on the Rune-Square. TLIA *will* work, though.
BCYG	
SCOH	This *won't* work in the Code-Square. It's family includes SBUP, BEPG, etc.
SMEA	Includes BORP, etc.
SOUR	Only one of these.
CTAY	Find three other relatives.
SCYR	Does *not* work in the Rune-Square but *does* in the Code-Square.
SPAH	It and its three relatives have no symmetry but are still "orderly arrangements." They are called "split-diagonals." They only work in the Rune-Square.
CLIY	Its relatives are only those in columns or rows 2 and 3. It only works in the Code-Square.

If students have omitted some of these in their searches, give hints to jump-start the search for more of them. NOTE: Students may try to list several more groups of four numbers that add to 34 but which have *no orderly arrangement or relatives.* LICY is one of many. The visual-spatial, and mathematical, challenge is to seek out only *orderly, symmetrical,* or *patterned* ones like those I've listed. Make it a whole class project to seek them all out!

The Rune Square and Games: Runes are fascinating mysterious writing used many centuries, even millennia, ago by Europeans. The *Lord of the Rings* Trilogy makes extensive references to them.

The Rune Game challenges students' design capabilities as well as their mental math. Encourage them to work with variations in their runes until one that is really satisfying comes up. There is no "right" or "wrong" way to create the runes. The only rule is that all pieces must be *connected* or inside others, and they must be *oriented* the same way as they are in the box they came from. Size is also of no concern — some may be made very large, some may be tiny.

A Box of Patterns: This activity creates a very colorful grid that becomes a good visual record of the patterns of 34-totals a group or individual has discovered. Often this will make the patterns more meaningful to some spatially-strong students.

A Design Surprise: This is yet another way to convert the magic squares to a visual-spatial form. The designs made are not simplistic or line-symmetrical and thus are hard to appreciate until they're colored. Groups of them can be tiled together on a bulletin board to create a splash of color.

Extension: Perfect Square. Students can explore why the Rune-Square is called a "Perfect Square." They need a sheet of large-square graph-paper (or the "Copycat" sheet from the "Symmetry" Activity 2). They will put the numbers from the Rune-Square into a square of boxes in the middle of the sheet. More Rune-Squares of numbers are made to join each of the four edges of the first square. Even the open corners are filled with four more Rune-Squares. This leaves the first square completely surrounded by others just like it. With a colored pen they outline any square of nine numbers they want in the whole array. It should check out to be a Magic Square.

The *Box of Patterns* Activity: This activity produces a very colorful shower of symbols that highlight the various spatial arrangements of 34-combinations in the Magic Square. Because it could become very tedious if done in one sitting by a lone worker, it makes a good group project. One way to be thorough is to start with one box at a time and color all of the configurations that involve it. Have a contest to see who can color the most symbols representing 34-totals.

 Make a Zillion Magic Squares. This has some interesting results about Magic Squares. Basically, squares remain magic if you interchange two *symmetrical* rows or columns, like the first and fourth or the second and third. They also remain magic if a number has been added to, subtracted from, multiplied by, or divided into, all elements.

Important Teacher Extension: This means you can choose a number like 1/4, or .022 and multiply it by **all** nine numbers of a 3x3 square or all 16 numbers of any 4-by-4, putting the result in each box. You will get a Magic Square requiring fraction or decimal addition to check it.

If students say, "No need to check, I'm sure it's magic," rig one box with a "**clinker**" by changing that box's number to something slightly different. Challenge them to find the "clinker" by finding which *row* and *column* do **not** add to the same total as all the others. The number at the *intersection* of this "bad" row and bad column is the "clinker."

5 by 5 Square: The 5-by-5 Magic Square has rows, columns, and diagonals that add up to 65. Students can check whether any other patterns add to 65. They won't find very much except that (9+ 14+ 19) + (11 + 12) and its three equivalents work. These are called "split diagonals. As shown, the square can be made by counting to 25 while writing groups of five numbers diagonally. Just jump each gray number into the blank postion in the magic square that has the same row or column as the original postion and as far away from it as possible.

Extension: Reverse Design. Explorers may want to try the **"Design Surprise"** in reverse. That is, make a grid of 16 dots numbered as before, draw four *symmetrical* line patterns on it, resembling ones in their designs, then translate these to rows of a square and see if it is magic. (Sometimes it will work and sometimes not.)

CLOCK PICTURES

If you want to tell time, it's best to have twelve numbers on a clock. But if you want to study the patterns that numbers make, use a **10-Clock**. It starts with a 0 at the top and goes to 9 as you go around. With the trusty 10-Clock you'll learn to understand the patterns that special groups of numbers make, because they'll draw interesting designs on the clock face.

There are several ways to get numbers that have a pattern to draw a picture on this clock face. Here is an example of a group of numbers that has a pattern:

| 4 8 12 16 20 24 28 32 36 40 44 48 |

Of course, these are the multiples of 4. These increase by 4 each step. Here are some ways these special numbers draw on the 10-Clock:

First way (end or units digit)

 a. Use the multiples of 4 and the clock at the right. Draw a line from the **end digit** of one number to the **end digit** of the next and so on. Use a ruler! Remember: the end digit is the **units** digit of the number.

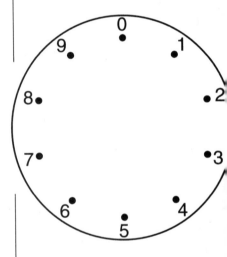

 Here's how to start: The end digit of a single digit number, like 4, is just 4. So, start at 4, make a straight line to 8, then to 2 (the end digit of 12), then to 6 (the end digit of 16), and so on.

 b. Describe your picture or design here using at least **8 words**:

_____

 c. Color the design in an interesting way!

 Try Other Multiples. Multiples of 5 or 9 are examples. Get a **10-Clock Sheet** and do end-digit clock pictures for them. Color your designs when you've made them. Label them with codes like "Mult 5, end".

Label each clock to remind you of what multiples you used, and put the word "End" in your label.

helpful   to do   calculate   write

10-Clock Sheet

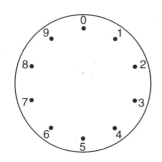

Label: _____

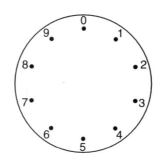

Label: _____

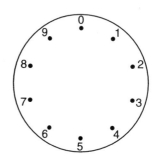

Label: _____

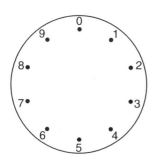

Label: _____

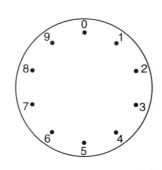

Label: _____

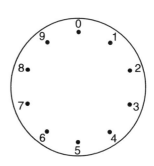

Label: _____

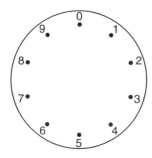

Label: _____

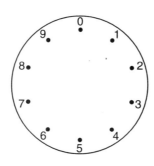

Label: _____

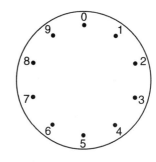

Label: _____

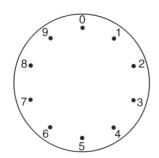

Label: _____

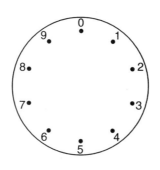

Label: _____

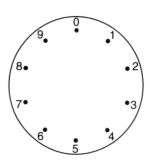

Label: _____

Second way (key digit)

Here's how to find a **key digit** for any number.
- Add up all the digits in that number to get a digit total.
- If the **digit total** has **two** digits, add those two together.

Example: Find the key digit for the number 377.
Add 3+7+7 = 17. Then add 1+7 = **8** the **key digit** for 377.

 d. Here again are the multiples of 4. Write their **key digits** in the blanks.

| 4 8 12 16 20 24 28 32 36 40 44 48 |

‒ — ‒ — ‒ — ‒ — ‒ — ‒ —

 e. Look at the key digits closely. Do you see a pattern in them? Look at **every other number**. Using a complete sentence, describe the pattern:_____

_____

_____

 f. Plot these key digits on the clock at the right by drawing a line from the first to the next (use a ruler!), and continuing until each key digit is connected in order.

 g. Carefully describe the picture or design you got using **at least 8 words:**_____

_____

 h. Color your design in an interesting way!

 Try Other Multiples. Multiples of 3 or 8 are examples. Use an extra **10-Clock Sheet** and do key-digit clock pictures for them. Label them with codes like "Mult 8, key".

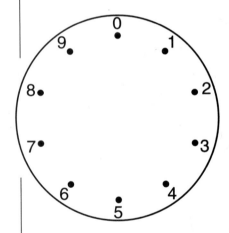

Label each clock to remind you of what multiples you used, and put the word "Key" in your label.

Third way (remainder number)

When you divide one number, like 17, by another number, like 7, you get an **answer 2** and a **remainder 3**. Here are some things to **watch out** for in finding remainders:
- Dividing 35 by 7 gives **answer** 5 and **remainder 0**.
- Dividing 5 by 7 gives **answer** 0 and **remainder 5**!

helpful to do calculate write

If you divide a whole group of numbers by **7**, you'll get a whole group of **remainders**. These can make a design on our 10-clock. Try this on the multiples of 4:

4 8 12 16 20 24 28 32 36 40 44 48

 i. Divide each of these by 7 and list the **remainders** here:

— — — — — — — — — — — —

 j. Draw line segments for them on the 10-clock at the right and label it *"Mult. 4, Div 7."*

 k. Color your picture in an interesting way!

 l. Describe here, **using at least 8 words**, the pattern you got:

_____

_____

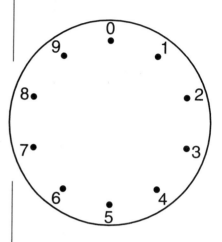

 Try Other Multiples. Divide them by 7. Use the extra **10-Clock Sheet** and make remainder pictures for them. Divide your multiples by 6 (or another number of your choice) and make even *more* remainder pictures on your 10-Clock Sheet.

Label each clock to remind you of what multiples you used, and put the word "Div 7" in your label. Use "Div 6" when you divide by 6, etc.

Clock Picture Research

You can do experiments just like a scientist with clocks in your laboratory. Each clock on the Clock Sheet has a blank line under it so you can label what the clock design shows. Here are some clock design research projects that will often surprise you. These will start you off then you can continue on your own.

Use as many 10-Clock Sheet pages as you need.

Multiples of 2

 a. Make a list of **multiples of 2** up to 12 x 2.

 b. Draw a 10-clock design for their **end digits.** Label it *"Mult. 2, End."*

 c. Draw another 10-clock design for their **key digits**. Label it *"Mult. 2, Key."*

Other multiples end digits and key digits

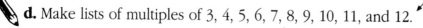 **d.** Make lists of multiples of 3, 4, 5, 6, 7, 8, 9, 10, 11, and 12.

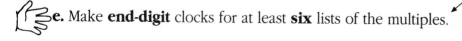 **e.** Make **end-digit** clocks for at least **six** lists of the multiples.

You can also find these in a times table.

Use the "Mult __, End" label for each clock as in the Multiples of 2 activity earlier.

helpful to do calculate write

f. Make **key-digit** clocks for at least six lists of the multiples.

Use the "Mult __, Key" **label** for each clock.

 g. Write **two or more complete sentences** on the back of your 10-Clock Sheet telling about the various clock designs you have found.

 As you write, mention such things as **likenesses**, **differences**, **weird features**, **your favorites**, and **patterns** you see in the numbers you connected.

Other multiples' remainders with 5

Use your lists of **multiples** of the numbers from **6** to **12**.

h. For **four** of the multiples lists make lists of **remainders** you get when dividing every number in each multiples list **by 5**.

 i. Make clock designs for each of your lists of remainders.

Label them "Mult 7, Div 5," "Mult 9, Div 5," etc.

j. Write two or more sentences on the back of your **10-Clock Sheet** describing your results.

See **g** above for suggestions for your writing.

You're the scientist

 k. As in the activities **h** through **j**, continue to explore possibilities from your multiples lists for remainder lists. Remember to label what you try on your **10-Clock Sheets**.

Fast growing numbers

Here's a new list of interesting numbers to make clock designs from:

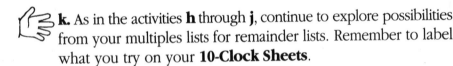

1 4 9 16 25 36 49 64 81 100 121 144 ___

Try to find some others after the blank, too!

Do you recognize them? What would be the next number in the sequence? Find out by subtracting 4 - 1, 9 - 4, 16 - 9, etc., all the way along to see a pattern.

Name the 50th. If you look at the numbers right and think mathematically, you can predict what the 50th number in the sequence will be without writing any more numbers! What is it? _____

l. Plot two 10-clock diagrams for these fast-growing numbers using the **End-Digit** and **Key-Digit** methods. Label them *"Fast, End"* and *"Fast, Key."*

 m. Plot **three** 10-clock diagrams using the **remainder** method. For one clock, divide each fast-growing number by **4.** Label it *"Fast, Div 4."*

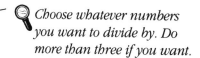 *Choose whatever numbers you want to divide by. Do more than three if you want.*

 n. Comment in at least **two sentences** on the back of your **10-Clock Sheet** about the patterns that the fast-growing numbers have made on the 10-clock.

*These numbers are usually called **"Square Numbers"** because any one of them is a number of counters that will arrange in a perfect **square**.*

Cube Numbers
There are other special numbers called "cube numbers." Their name means that each one is a number of **small** cubes that would stack to make a perfect **large** cube shape.

Here they are:

The first in the number list is 1 x 1 x 1, the second is 2 x 2 x 2, and so on, up to 10 x 10 x 10.

1 8 27 64 125 216 343 512 729 1000

Try stacking 27 cubes and 64 cubes into perfect cube shapes.

 o. Plot clock designs for the cube numbers using **End-digit, Key-Digit,** and some **Remainders**. Label them *"____, Cube."*

 p. Comment in at least **two sentences** on the back of your **10-Clock Sheet** about the patterns that the **cube numbers** have made on the 10-clock.

 The Doubling Numbers. Try a bunch of 10-clocks for the sequence **1, 2, 4, 8, 16, 32, ...**

 Nine's Fingerprint. Multiply any 3 or 4-digit number, by **9.** Find the **key-digit** for the answer. Do this for **five** other such numbers multiplied by 9. What do you notice about their **key digits**? Check out the key digits after you multiply any **five** numbers by **3**. What do you notice?

 Musical Patterns. Tape the numbers 0-9 on 10 of your favorite keys on a piano or xylophone. Make your 10-clock designs musical by playing the notes of digits in the order that they are connected in the clock design. Play the cycle of numbers over and over until you can hear its distinctive tune. Compare design-tunes and pick some of the most pleasing ones.

Or you can use 10 identical bottles filled to different levels with water.

Change labels on the keys until you get some really good tunes from some clock designs.

helpful to do calculate write

People Patterns (for a group). Find **ten** students to stand in a circle, each with one of the numbers 0 through 9 taped or pinned to the front of his or her clothes. You now have a **living 10-clock. Colored string or yarn** can join numbers across the circle in the same way that you drew lines before. Try "drawing" with string some end digit, key digit, and remainder sequences and see how they look as **"people-clock designs."**

The Twelve Clock

For this activity you get to use a **12-Clock Sheet**. Get one from your teacher.

q. With your ruler join all number-pairs on a 12-clock face that **add to 12** .

That is, draw a line between 1 and 11, 2 and 10, etc., until all such pairs are joined.

r. On more 12-clocks draw lines connecting pairs that **add to 11.** Do separate clocks for those that add to **10**, **9**, **8**, **7** and **6**.

s. Describe, in complete sentences, on the back of the **12-Clock Sheet** how the lines change from clock to clock.

t. On a 12-clock face connect pairs that **subtract** to make 6, like 7 and 1. On more 12-clock faces connect numbers making a 5-difference, a 4-difference and so on down to a 2-difference.

u. Describe, in complete sentences, on the back of the **12-Clock Sheet** how the pattern changes from clock to clock.

12-Clock Sheet

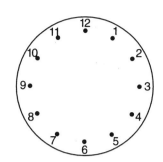

Label: _____

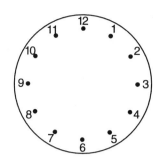

Label: _____

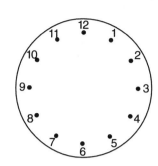

Label: _____

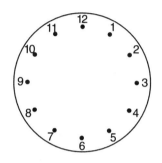

Label: _____

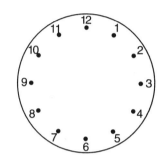

Label: _____

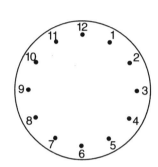

Label: _____

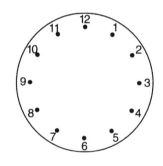

Label: _____

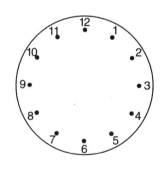

Label: _____

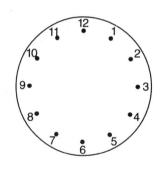

Label: _____

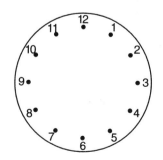

Label: _____

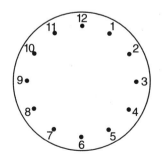

Label: _____

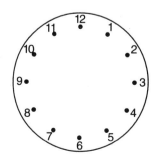

Label: _____

Teacher Notes

Intelligences Emphasized: Visual-Spatial, Linguistic, Musical, Bodily-Kinesthetic, Naturalist (sorting and classifying patterns)

Math Skills, Concepts: Multiples (times tables), division, remainders, squares, cubes, patterns, digits of a number, mental addition, symmetry

Materials: Ruler, pencil, lined paper, colored pens, 10-Clock Sheets, 12-Clock Sheets, piano or xylophone

Level of Challenge: Moderately easy to moderately challenging

Overview: This spatial activity has an exploratory feel and a sense of surprise as designs occur. Students must constantly use mental math and turn answers into designs. The Activity also regularly encourages the student to explore and describe the designs, both orally and on paper. There is almost no end to the number of designs that can be extracted from numerous sets of orderly numbers and the clock-drawing techniques. The payoff is increased intuition about hidden, ordered, cyclically repeating behaviors in familiar number sequences.

Introduction to the Class: Comment on how surprising and patterned numbers can be. Ask the students if they recall any examples of this. (The "Fair and Square" Activity is a good example if they have done it, or the well-known pattern of the 9's table — see Chapter 9 — is another good example.)

Justify the Activity with the idea that the ability to recognize number patterns is an important part of being quick with numbers, and this recognition gets better with training. Patterns also underlie work in *algebra*, since algebra is a symbol system that generalizes and symbolizes patterns in the behaviors of numbers.

Give an example: 8, 9, 6, 7, 4, 5, 2, 3, 0, 1. Ask for suggestions about how it is patterned and encourage students to use good grammar in their oral descriptions. If they don't see much, suggest looking at every other one, or reading them backwards. Emphasize that their skill with these will improve in the Activity, which taps students' spatial intelligence to foster deeper understanding and appreciation of numbers. Start with pages 5-1 and 5-2 only. Pages 5-3 and 5-4 are another block of work.

jottings

Answers with Comments on the Tasks

You can bring out much more understanding if you *encourage students to look for and discuss the patterns in the end-digit, key-digit and remainder lists that become clock pictures.* That is, studying the lists from a *numerical* standpoint and a *spatial* standpoint is best.

For instance, the key-digit pattern of multiples of 7 is 7, 5, 3, 1, 8, 6, 4, 2, 9, 7, 5, 3. These look just like a jumble of numbers to some students, yet they have a fascinating internal order. Reading the numbers backwards is very revealing of a "neat" pattern of reverse odds, reverse evens. Follow this by tracking how the clock pattern reflects and reveals this order.

Encourage the *labeling* of each clock made. Run off *extra* **10-Clock Sheets** so students will feel free to explore many clock pictures. Encourage and model the construction of *good, clear sentences* describing results obtained with the 10-clocks — this is valuable composition work. Groups can pool their remarks about the designs and come up with a "definitive" group statement about them using "edited" sentences. This is the same kind of thought process a biologist or geologist uses in sorting and classifying natural forms.

Clock Picture Research

Fast Growing Numbers. First are the square numbers. With the next one being 13 x 13 = 169 and the 50th one being 50 x 50 = 2500. A discussion of where they appear in a multiplication table, (diagonal) and how they relate to the area of a square, or to any square arrangement of counters or cubes would be valuable. Later in math, squares will be used in finding areas of circles and in using the Pythagorean theorem. Their end-and key-digit designs are interesting, as are those of the cube numbers.

 The Doubling numbers; Nine's Fingerprint The doubleing numbers' end-and key-digit designs make a nice "bowtie with loose thread" and a "pair of wings" that retrace themselves. Multiplying *any* number by 9 makes the answer's key-number always 9, while the key-number is a multiple of 3, 6, or 9 after multiplying *any* number by 3.

***The Twelve Clock.* r, s.** In the 12-Clocks there are many parallel-line patterns that result with sums. **t, u.** The subtractions result in line pictures that keep rotating around part of the clock.

6 PEOPLE AND MORE PEOPLE

The Population Facts

Is our planet very crowded with people or not? What do you think? Write the first answer that comes to your mind here:

_____

The activities you are about to do will help you think more about your answer.

a. Here's your first mission to prove you are ready to be given some important information about your planet. You need to be sharp on how **millions** relate to a **billion** (1000 million), so read this box **carefully**:

> Every 100 million is 1/10 billion (or .1 billion).
> Every 10 million is 1/100 billion or .01 billion.
> Add them: 110 million is .11 billion (or .110 billion).
> So 256 million (256,000,000) is .256 billion.

b. Got that? To prove you did, answer these questions by filling the blanks with a **decimal** or **whole number**, then check them with a partner and your teacher:

1) 13 million = _____ billion
2) 8,000,000 = _____ billion
3) 849 million = _____ billion
4) 2 billion, 64 million = _____ billion
5) .13 billion = _____ million
6) 3.007 billion = _____ billion, _____ million
7) 6 billion + 37 million = _____ billion
8) 2.065 billion = _____ billion, _____ million

Now you're ready for the information about your planet. You'll be using it to make an amazing graph of the earth's population for the years 1950 to 2025 AD.

c. First just **read** them and think of what they mean:
- In 1950, the world population was 2.513 billion.
- In 1960, it was 514 million more.
- In 1970 it was 3,678,000,000. By 1975 it had risen 376 million more.
- In 1980 it had reached 4.478 billion.

The facts came from several sources, so they're said in different ways.

helpful   to do   calculate   write

- In 1985 there were 387,000,000 more people.
- In 1990 the world population was 5,292,000,000.
- In 1997, it was 5.919 billion.
- **Each year** until 2000 the world population will increase by **another England and another Spain** — about 97 million people! After 2000, early years add about 105 million (a France + a Korea) to the earth.
- India is the **second largest country** (next to China) and its population is 970,000,000, or very close to **one billion**. At the rate our earth's population is growing, we will have gained **another India** amount of people just between **1990** and **2000**!

 d. Figure out from the facts the population of the earth for **2000 AD.** _____ **2008 AD.** _____

A Population Graph ·

Here are some steps for making a good world population graph. ◄——

Use the **World Graphing Sheet** and a **ruler**.

 a. Go through all the population numbers above and write them as **billions** (like 5.919 billion) for each year. ◄——

Some are already written that way.

 b. On the World Graphing Sheet, the heavy **horizontal** line at the bottom will be the **time axis** for the years from 1950 to 2025. Number in three year jumps at the "**tic marks**" (see figure).

Write "Years" once in the space below the numbers.

 c. The heavy **vertical** line near the left edge will be the **population axis** with numbers from **2.5** to **6.3** billion, in jumps of **.1** billion at each tic mark.

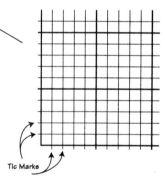

Tic Marks

 d. Carefully plot a "**population point**" for each of the years of your list. Find it by going **straight up** from the year and **straight over** from the population number until they meet at a point. Then join your point to the last one with a crisp **ruler** line.

*Then write "**Billions**" vertically along the paper's left edge near the numbers.*

Questions and Activities Involving Your Graph

 e. The point-to-point line segments form a curve. Is it getting **less steep** or **steeper** as it goes? _____ ◄——

Remember, it will still curve past the top of the graph!

f. Figure out from your graph The world population in 1955 _____; in 1965 _____; in 2002 _____.

helpful to do calculate write

World Graphing Sheet

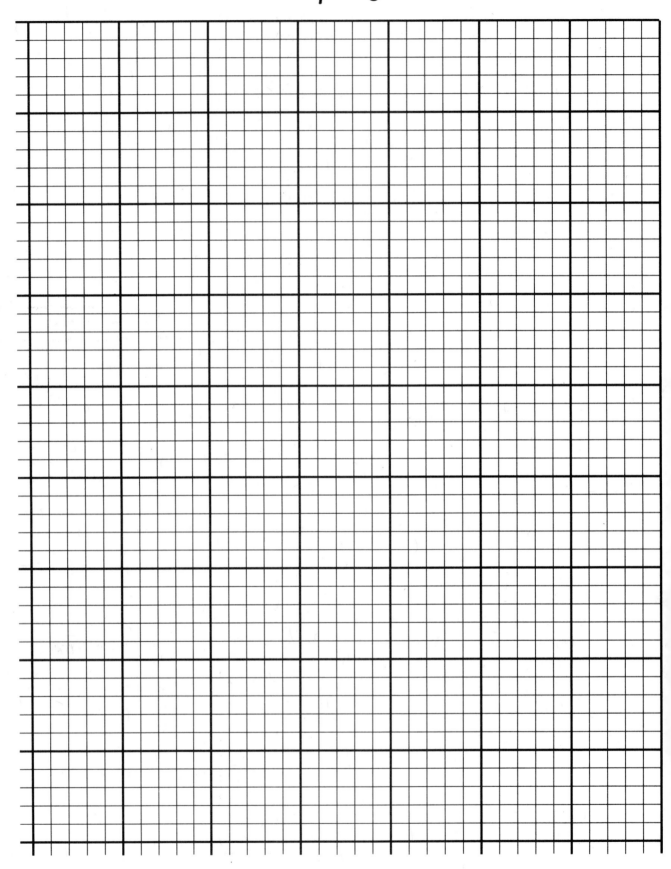

g. Find **another** sheet of paper and tape it so its bottom touches the **top line** of your World Graphing Sheet. With a **ruler** extend the **vertical** line for the year 2025 **straight up** onto your new sheet. **Continue the curve** of your population graph until it hits the 2025 line. Predict from your extended graph what the population in 2025 will be. Write it here: _____.

 h. Your prediction for 2025 is based on **past** population growth. Many experts feel this prediction could be **too large.** Write here what happenings in the world could cause the curve not to rise so fast: _____

_____

_____

 i. A counter-graph. Find **100** coins, cubes, or chips as counters that are in at least **three colors.**

    1) Make a **solid square** of 25 counters. It represents the 2.5 billion people in 1950. (Each counter means .1 billion or 100,000,000 people.)

*Use **one color**, and **don't re-use** this color until step **5)**.*

    2) Using a **different** color, add on counters **in rows of five** to show the population change to **1960.**

    3) Change color and add counters for the change to **1970.**

    4) Continue to **1980**, **1990**, and **1997**, changing the color each time, until you reach **2000.**

*Keep only **five** in each new row.*

    5) Then change to the color used in step **1)** and put on the amount that represents the growth to **2025.**

Still in rows of five.

    6) How does step **5)** compare in size with the original 1950 square?

_____

    7) Read the box to help your Counter Graph amaze you, then **discuss your insights** with partners.

***Write** or verbally **report** your comparisons.*

DON'T FORGET: The **1950 square** represents what the earth's population became in the **many thousands of years** since the **first persons** walked the earth!

The rows from **1950 to 1997** are only **47 years** of population growth! Compare this with the 1950 square.

The rows you added from **1997 to 2025** are only **28 years** of population growth. Compare these with the 1950 square!

World Meeting

Imagine that the entire population of the world in 2000 decided to meet **all in one place (!),** and you were invited as a **speaker** to present **three** ideas at the meeting. Pretend that you would have no "stage fright."

 a. Carefully and seriously think of what **three important ideas** you feel would help the **whole world**. Discuss your thoughts with others. Write them here (or on the back of this page):

1._____

2._____

3._____

Question: How much **room** would this world meeting take?!!

 b. Below are pictures of four U.S. states ranging from the smallest to the largest. Color the meeting area you think would be needed in one or more of these states. Just take a guess! ◄—

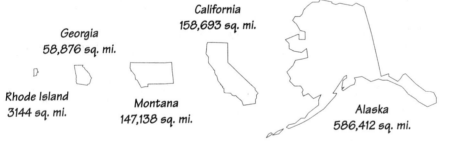

Georgia
58,876 sq. mi.

California
158,693 sq. mi.

Rhode Island
3144 sq. mi.

Montana
147,138 sq. mi.

Alaska
586,412 sq. mi.

*Assume each person gets **one square yard** to stand on.*

 c. Calculate how much area in **square miles** the meeting would ◄— take: _____ (See the hints/steps below.)

 Each person (man, woman, or child) gets to stand and put their stuff on **one square yard** of ground. Since **each** square yard is **9 square feet**, calculate how many **square feet** are needed by all the world's people at the meeting:

_____

 A **square mile** is 5,280 by 5,280 **square feet**. How many **square feet** are in one **square mile?** _____

 Use the two results above to figure how many **square miles** ► the people need.

 d. With a different color **fill in the area** needed for the meeting in one or more of the states above and **compare** with your estimate.

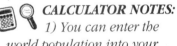

CALCULATOR NOTES:
1) You can enter the world population into your calculator by leaving off the last six zeros. After you get your answer, move your decimal point six places to the right and it will come back to the correct size.

2) When you divide $8000 \div 2000 = 4$, you could get the same answer by dividing $8 \div 2 = 4$. Remove the same number of zeros from both numbers you're dividing, and your answer will still be right!

helpful   to do   calculate   write

Ring Around The Earth

It's **24,894 miles** around the earth at its **fat** equator. People at the poles are **119 miles closer** to the center of the earth than they are at the equator!

 a. Why do you think the earth is fatter at the equator?

_____

 b. If all the people in the world in 2000 held hands, could they reach around the equator? _____ If so, **how many times** around the equator could they reach? _____

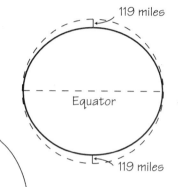

 Assume each person's hands can stretch an average of **four feet**. Remember there are **5,280** feet in a mile. Think carefully when to multiply and when to divide. Round answers to **one decimal place**.

 Class Reach. Predict how far all the students in your class can reach with hands joined then try it and measure! You can even have a contest with others to see who has the closest estimate!

 Local Reach. Do some research on population around you, then predict how far all the people in your town, city, rural area, or state could reach while joining hands.

 Moon Reach. It is 238,000 miles (on average) to the **moon** from the earth. If our hand-holding line of all people on earth could stretch out in space toward the moon, what **part** of the earth-moon distance, or **how many times** the distance, would the line reach? _____

 Show your answer on the picture.

 CALCULATOR NOTE:
Remember to leave <u>six</u> <u>zeros</u> off the number of people in the world when you multiply with the calculator. Do your operations. <u>Then</u> you can replace six zeros at the end <u>or</u> move the point six decimal places to the right, and your answer will be right.

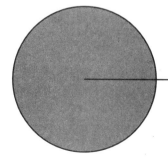

EARTH

238,000 Miles

MOON

helpful   to do   calculate   write

Do the **World Meeting, Ring Around the Earth, and Moon Reach** activities again, but use the earth's 2025 AD population.

 Is the Earth's Population Large or Small? Review your results from all of these activity pages. Discuss in your group how each one made you feel about the present and future size of the population of the earth.

*To get ideas, look back at your results from: **The Population Facts**, **A Population Graph**, your **Counter Graph**, **World Meeting**, **Ring Around The Earth**, and the **Moon Reach** activity.*

In the space below, either
1) Write some paragraphs containing your thoughts and feelings about the size of the earth's population now and in the future, or
2) Draw one or more pictures that show your impressions about the earth's present and future population size.

helpful

to do

calculate

write

Teacher Notes

Intelligences Emphasized: Visual-spatial, Intrapersonal, Interpersonal, Linguistic, Kinesthetic

Math Skills, Concepts: Working with large numbers, naming and rounding decimals, calculator skills, graphing, estimation, area, unit conversions, rates, circumference

Materials: Calculator, pencil, World Graphing Sheets, ruler (tape measure), colored chips or cubes

Level of Challenge: A good challenge for intermediate grades

Overview: This activity highlights real life information (world population, sizes of states, size of the earth) as well as reflection on the rapid growth of our population. Students are challenged to make such information more digestible with appropriate mathematical tools. A discussion of the of the reality and implications of earth's rapidly expanding population would be a great follow-up.

Note that the Activity "Addresses and Distortions" makes a good companion for this Activity because it also develops the graphing concept, though in a very different way. Either one can be done before the other, depending on your instructional needs.

Introduction to Class: A more in-depth discussion of the students' initial notions of population on the earth would add relevance to the Activity they are about to begin. Some may have heard an estimate of the number of people that are now on the earth, but don't dwell on this number — it will emerge in the Activity. It could also be of interest to discuss what it would be like to have a meeting of all the people on earth in one place (without hinting in any way how much room this meeting would take).

Some practice on billion=1000 million, decimal parts of billions and millions, etc., is best before handing out the sheets, then reinforce the same material on 6-1. Depending on the students' previous experience, a short discussion of line-graphs and their purpose would also be worthwhile.

Then simply distribute pages 6-1 through 6-4, either to work on in groups or as individuals. The rest of the activities can be done in later sessions.

Answers

The Population Facts. a. 1) .013; **2)** .008; **3)** .849; **4)** 2.064; **5)** 130; **6)** 3, 7; **7)** 6.037; **8)** 2, 65; **d.** 6.21 billion in 2000 AD; about 7.1 billion in 2005 AD (about a 1.7% growth rate).

A Population Graph. A graph requires two sets of axes for showing the relationship of two changing quantities, which, in this case, are years vs. population. The Activity guides students in making their own graph of the population figures. They will get a curve similar to the one shown at the right.

Extension: Before 1950. A further intriguing question is, "What did the curve look like *before* 1950?" The further back in time it goes, the more horizontal it appears and the lower it gets. Research on the exponential curve or on early earth populations would make a good project.

Questions and Activities Involving Your Graph

e. The graph is getting steeper as it goes.

f. 1955: about 2.8 billion; 1965: about 3.3 billion; 2002: about 6.6 billion.

g. The extension of the graph line, if done correctly *with the curve growing steeper*, will reveal a 2025 population of about 9.5 billion(!)

h. This is a complex question, though student answers may be simple. The actual United Nations projections for the 2025 population state an interval of 7.9 to 9.1 billion (the average being 8.5), which is *lower* than the extension of this curve.

Here's why. It's predicted that there will be some slowing in the rate of growth from the present 1.7% a year to 1.4% a year. To grasp the effect of this predicted slowing, picture a car which accelerates on the highway until it exceeds the speed limit. Now imagine that it is still exceeding the speed limit and is still accelerating, but it's acceleration is a bit less — it will still reach very unsafe speeds, but a little bit later!

The UN's lessening of the acceleration of the future population curve relates to a foreseen reduction of birth rates in some developing countries as well as in Europe and the U.S. It might also reflect life loss from future famines, regional wars, and epidemics. Wars often occur when countries continue to intensify their competition for limited water and agricultural land resources.

i. This activity visually/spatially/kinesthetically dramatizes the growth. The 1950 population square begins to look very small

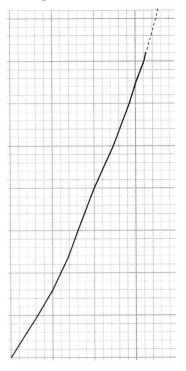

next to the amount added on by 1997. It is also outdone by the 28-year amount accumulated from 1997 to 2025. **7)** Encourage careful reflection and discussion about what the counter graph is revealing as pointed out in the **DON'T FORGET** box.

World Meeting

a. This *intrapersonal* activity will help the students to emotionally relate to the math exercise they are about to do, as well as clarify their values concerning the world or life. A discussion in the whole class, as well as in groups, before writing would be a way to deepen this exercise.

b. Most will overestimate the amount of room in the states shown that will be needed for the meeting, but make no comments in advance that will change their estimates.

c. A good kinesthetic and spatial connection is to mark or tape a square yard with its nine square feet on the floor. Better yet, mark a few (like nine of them) in a large square and have students stand in each to see how realistic this assumption is.

Students get confused about whether to divide or multiply the numbers involved. Get them going with comments like, "If **each** person requires 9 sq. ft. to stand on, how many sq. ft. will three people need? Four? One hundred? etc."

As indicated in the CALCULATOR NOTE, students won't get the full number for the earth's population, 6,210,000,000 onto the screens of their calculators. Make sure they follow its recommendation, and spend time separately on this valuable calculator finesse so that all understand why they are doing it. Leaving off six zeros is equivalent to **dividing** by 1,000,000. Their whole answer will end up being 1,000,000 times **too small**. Thus, moving the point six places to the **right** at the end will make it grow (**multiply**) by 1,000,000 again.

- If students are stuck on how many **square feet** are in a **square mile**, you can prompt them with, "If we put 5280 sq. ft. along a mile-long line, then another 5280 sq.ft. next to those and another 5280 next to those until we had placed 5280 lines of 5280 sq. ft., these would fill a square mile." [27,878,400 sq.ft.]
- Another prompt for those stuck: "The number of square feet required by everyone must be parceled out in bundles of 27,878,400 sq.ft. that fill a square mile. Which operation parcels out (divides up) numbers? **Divide**, of course."

• The answer is a surprising **2004 square miles** (rounded off) or only about 64% or 2/3 of tiny Rhode Island. They need color only a dot in each state!

A **discussion** about whether this means the earth's population is big or small for the earth should bring up many ideas. This discussion is appropriate after any of the subactivities. It will be summed up in the "**Is the Earth's Population Large or Small?**" activity below. Of course, the big problem is that each human being requires *more* than a square yard and many want *lots* of land. Agriculture, cities, highways, and industries gobble up land. Following the discussion, written statements of students' opinions based on the numerical results would be an excellent assignment.

Ring around the Earth. This activity is designed to focus attention on the globe. **a.** The "fat earth" question makes a good science discussion. The earth would feel like a tomato if it were a size to hold in your hand, so it's a bit squishy. The spinning causes the Equator to squish outward more than the poles.

b. The CALCULATOR NOTE is valuable here as previously. Operation prompts for those stuck: "Each person requires 4 feet, so how many do 15 persons require? That's right, multiply the population by 4. Now parcel out these feet in bundles of 5280 (i.e., divide by 5280) to see how many miles they take. Parcel these miles in bundles of 24,894 miles (divide by 24,894) to see how many *earth circumferences* they will take up. [189]

 Moon Reach. This only involves taking the total miles the hand-holding earth population can reach, 4,484,090 miles, parceling (dividing) these in bundles of 238,000 miles to get about *20* times the earth-moon distance! They should show all these earth-moon trips on the picture.

 Is the Earth's Population Large or Small? This is a kind of culminating intrapersonal activity for the whole multi-faceted Activity. Each sub-activity generates various perceptions and feelings about the earth's population and this is a chance to draw them all together.

7 ADDRESSES & DISTORTIONS

Have you ever heard of an address of a building given as "the corner of 5th and Main?" Imagine a place where the address of any building, or your friend's home, is all numbers, like "the corner of 8th and 12th."

There's a foolproof way to set up a system like this. **Pilots** and **sailors** like to use it. They think of the whole earth as a kind of "city" with every spot having a number address they can go right to. They know exactly where they are all along the way, too.

First you'll learn to use something like their system; you'll make pictures from addresses, and then you'll learn to "warp" or "distort" those pictures with math.

This system was invented in France in 1630 by Mr. René Descartes (pronounced day-cart).

Later you can learn how pilots make addresses for any place on earth.

Landing at Addresses

At the right is the simplest type of the address system — two lines at a 90° angle Where they meet, the crossing point, is called the "origin." Always stand first at this "starting point" to find addresses. Think of it as "zero."

 There are ten equally-spaced points along the bottom line segment.

 Number the points 1 to 10.

 The numbers can mean any units you'd like - inches, feet, degrees, hours, years, cm, km, etc.

 Choose a unit you like and write it once on the line under your numbers.

 Number the ten points up the vertical line, starting from one. To the left of those numbers write another unit name you like.

ADDRESS SYSTEM

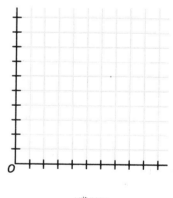

unit name

unit name

Each line is called an "axis" and both together are called "axes" (pronounced axeez).

helpful to do calculate write

(7, 3) is called an "ordered pair." Use this code to understand it: **(across, up) or (→ ,↑).**

The 7 and the 3 are called "coordinates" of a point. Together they make an "ordered pair" which is the point's address.

 To place **(7, 3)** on the "city" you numbered, stand at **zero,** go **across** → **7** steps to the right, then go **up** ↑ **3** steps and make a dot. You've found (7,3).

 On the first page, mark these eight points: (1, 1), (2, 3), (3, 4), (5, 5), (7, 3), (9, 4), (9, 7), (5, 5)

Connect the dots in order and get a "sky-picture".

Distortion Activities

Get a ruler and a Graphing Sheet from your teacher. The Graphing Sheet has each axis marked with 32 or 36 units.

Draw all requested lines **with a ruler only.**

 Mark on the Graphing Sheet the points of the **Starter Picture** listed below. Then join them with **pencil** line segments in order. The picture you get has something to do with math.

Remember the ruler!

Starter Picture: | **(1,3)   (3,3)   (4,7)   (5,5)   (7,7)   (7,3)   (9,3)**

 You're going to change and distort the picture you drew, kind of like using Mole's carnival mirror. You'll do that by changing the *address of each point* of the **Starter Picture** the same way, with a Distortion Rule. Then you'll plot the new points again and connect them with line segments.

Here are some Distortion Rules to try:

Distortion Rule 1: | In *every* **Starter Picture** pair **add 7** to the **first** number and **add 6** to the **second** number.

 a. Before you do this, predict or guess how this will change or distort your original figure:

_____

 b. Fill in the new addresses using the distortion rule:

(8, 9) (,   ) (,   ) (,   ) (,   ) (,   ) (,   )

 c. Plot them and use **green** line segments to join them in order.

 helpful
 to do
 calculate
 write

GRAPHING SHEET

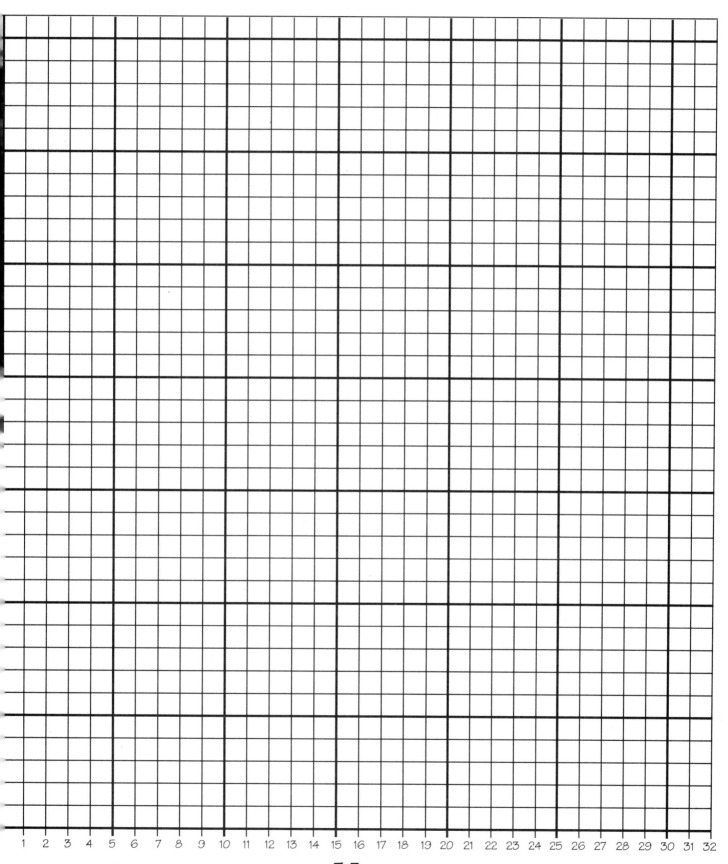

1 2 3 4 5 6 7 8 9 10 11 12 13 14 15 16 17 18 19 20 21 22 23 24 25 26 27 28 29 30 31 32

d. Write, in a complete sentence, how the picture has changed:

_____

_____

Distortion Rule 2: In every address of the **_Starter Picture_ reverse the numbers**, that is, put the second number where the first is, and the first where the second is.

a. Before you do this, predict or guess how this will change or distort your original figure:

_____

b. Fill in the addresses using the distortion rule:

(3, 1) (,) (,) (,) (,) (,) (3, 9)

c. Plot and join them with **red** segments.

d. Write, in a complete sentence, how the picture has changed:

_____

_____

Distortion Rule 3: For every number pair of the **_Starter Picture_**, **add** the numbers of the pair together. Put that **sum** in the **second place** of the new pair. Keep the **same first number** in the first place of the new pair.

Example: In (1, 3), add 3 + 1 = 4, so your new pair is this: (, 4). Put the 1 back in the first place to get (1, 4) to plot.

a. Before you plot these, predict or guess how the rule will distort your original figure: _____

The first two are done for you.

_____

b. Fill in the new addresses using the distortion rule:

(1, 4) (,) (,) (,) (,) (,) (9, 12)

c. Plot the points and connect with **orange.**

d. Write, in a complete sentence, how the picture has changed:

_____

_____

helpful | to do | calculate | write

Distortion Rule 4: | Do the same steps as **Distortion Rule 3**, but put the **sum** in the **first** place and keep the **second** number the **same**.

 a. Predict or guess how this will distort your original figure

_____

 b. Fill in the addresses using the distortion rule

(4, 3) (6, 3) (,) (,) (,) (,) (12, 3)

 c. Plot and join them with **purple** segments.

 d. Write, in a complete sentence, how the picture has changed:

_____

_____

Distortion Rule 5: | **Multiply** the first **and** second coordinate numbers in each address of the **Starter Picture by 2**, then **add 14** to the **first** coordinate

 a. Before you do this, predict or guess how this will distort your original figure:

_____

 b. Fill in the addresses using the distortion rule:
(16, 6) (,) (,) (,) (,) (,) (32, 6)

 c. Plot and join them with **brown** segments

 d. Write, in a complete sentence, how the picture has changed:

_____

_____

Distortion Rule 6: | **Multiply** the **first** coordinate **by 3** and the **second by 4** in all pairs of the **Starter Picture.**

 a. Before you do this, predict or guess how this will distort your original figure:

_____

 b. Fill in the addresses using the di,stortion rule:
(3, 12) (,) (,) (,) (,) (,) (27, 12)

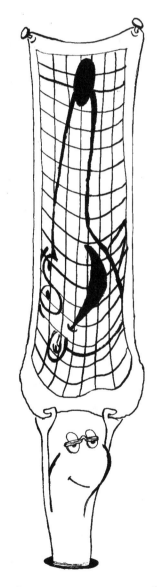

 helpful to do calculate write

c. Plot and join them with **yellow** segments.
Write, in a complete sentence, how the picture has changed:

_____

_____

> Make up another **Distortion Rule** for
> the addresses of the **_Starter Picture_**.

 a. Write the rule here: _____

_____

 b. Predict or guess how your rule will change the picture:

_____

(,) (,) (,) (,) (,) (,) (,)

 c. Plot and connect in **turquoise**; see what happens.

 d. Write, in a complete sentence, how the picture has changed,
and how close your prediction was.

_____

_____

 Make More Rules. Get another Graphing Sheet from your
teacher. Make up more rules for changing the **_Starter Picture_**.
Try some unusual ideas for what to do to **every pair**.

 Make A Picture. Make up your own **_Starter Picture_** and
distort it with your own fancy rules.

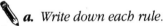 *a. Write down each rule.*

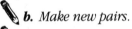 *b. Make new pairs.*

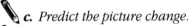 *c. Predict the picture change.*

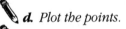 *d. Plot the points.*

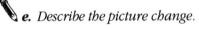 *e. Describe the picture change.*

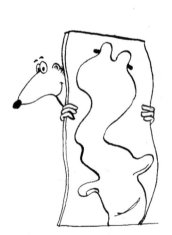

 helpful to do calculate write

Negative Coordinates

On a special "Negative Coordinate Graphing Sheet" is a set of axes that are numbered in both positive and negative numbers. Negative coordinates are to the **left** on the horizontal line and **below** on the vertical line.

Here's a new **Starter Picture**. Connect the dots with pencil segments (using a ruler) as you plot them.

Starter Picture II: (1, -2) (0, -3) (-1, -4) (-2, -7) (-5, -5) (-7, -3) (-6, -1) (-8, 0) (-7, 2) (-6, 4) (-5, 6) (-3, 7) (-1, 6) (1, 6) (3, 5) (4, 4) (3, 3) (2, 2) (5, 1) (6, 0) (6, -1) (1, -2)

Here are some ways to change the addresses in the **Starter Picture**:

Distortion Rule a:	**Add 8** to each **first** coordinate, and **-10** to each **second**.

Distortion Rule b:	Take each pair from Rule a and change **each coordinate** in the pairs to its **opposite**, i.e., take (8, -9) and make it (-8, 9).

Distortion Rule c:	Subtract 8 from each **first coordinate** of each pair and **double** each **second coordinate**.

 Longitude and Latitude.

Discuss, research and ask your teacher for ideas on these:

 How is what you have done similar to what a pilot must do to find a particular place on the earth?

 How is the work you have done related to the concepts of longitude and latitude for locating places on the earth?

 Look up "longitude" and "latitude" in an encyclopedia or book and discuss the 🔍 questions with others. Write your answers.

---

*Remember to start at 0, move **across** for the first number, then **up-down** for the second. (-2, -5) is plotted for you.*

It uses the eyes, nose, and mouth on the Negative Coordinate Graphing sheet.

Write your new list first, or you'll get confused when you plot.

Remember, subtracting 7 from -2 makes -9 ! Subtracting positive makes a negative get more negative.

*Plot each picture in a new color. Write how the **Starter Picture** has changed. Add your own eyes, nose and mouth!*

Your teacher has information about this.

helpful | to do | calculate | write

NEGATIVE COORDINATES
GRAPHING SHEET

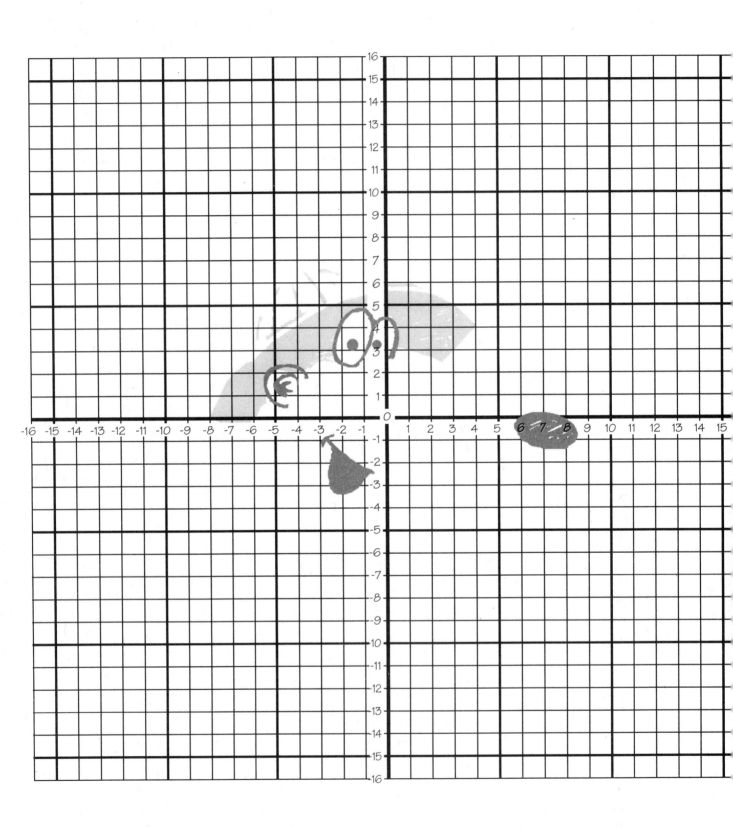

Teacher Notes

Intelligences Emphasized: Spatial, Linguistic, Bodily-Kinesthetic, Interpersonal

Math Skills, Concepts: Graphing, transformation of coordinates, latitude and longitude (optional), addition and subtraction with negative integers (optional)

Materials: Graphing Sheets (photocopied from these pages), African map (photocopied from the "Count on Africa" Activity)

Level of Challenge: Moderately easy to moderately challenging

Overview: This Activity is intended to introduce and apply the concept of **ordered pairs** that specify the two coordinates of a location. The Activity also develops in a very simple way a sophisticated idea (known in higher math as a mapping or transformation) which involves applying a fixed rule (function) to a set of pairs to distort the picture that they make on coordinate axes. Teachers who wish to go further are shown how to extend all of the activities to the negative coordinate numbers extending to the left and below the origin. Finally, in a short development that can be easily extended, comes the notion of longitude and latitude coordinates on the globe for locating any point in the world, and how it relates to plotting on regular Cartesian coordinates.

Introduction to Class: Before handing out the activity sheets, build interest in the concept of addresses by expansion of the brief reference at the beginning of the Activity to the way buildings in a city are located by cross-streets. Discuss how people generally describe locations to each other. Further interest and relevance can be generated with a guided "inner movie", sketched below. You can first post and point to Niger on the map of Africa mentioned under **Materials** above. Here's a sketch of how the imagination exercise might be conducted:

Students put their heads on their desks, or just close their eyes. You start the story right away or, for greater immersion in it, you lead a brief relaxation exercise in which they tune into each area of the body from feet to head to relax it, then they observe their own breathing, without forcing a change, to see if it is deep or shallow. After the relaxation, begin simply and slowly in a calm voice to set a scene something like this:

> *"Imagine you are flying a small plane over the African savanna in the east of the country of Niger. Hear the wind whistling past and the plane's loud engine. Below are endless*

Some students may already be comfortable with the rudiments of coordinates from playing "Battleship."

You can make reference to the fact that this system of addresses has been named "Cartesian Coordinates" after the French mathematician René Descartes who invented it in 1630.

*See also the reference to **"internal imagery"** in the spatial-intelligence section of Chapter 3: Seasoning Math with MI.*

Feel free to embellish this story with your talents.

*dry, grassy expanses and herds of grazing animals. You radio
to a friend at an airport about two hundred fifty miles away
in Zinder, the closest city, that you are landing. You do so
and walk around. There you are in the middle of a huge,
dry, brushy area with no mountains as far as the eye can
see. You explore then you spot a reddish sparkle. You clear
the dirt away, and before you glows a giant, deep crimson
jewel, as big as a basketball, perhaps left by thieves or the
French Foreign Legion forty years ago. It looks rare and
incredibly valuable. You want to take it with you. Uh, oh,
there's a catch. It is completely set in a socket of iron which
is anchored into rock with some kind of solid cement!*

*"You have no sledge hammer or tools to free it, and you
have nothing large to mark the location. You need to fly out
to get help, or at least to get tools. And, of course, you want
to remember exactly where this treasure is! There are no
major landmarks around! You are hundreds of miles from
a city! There could be a wind storm that covers the jewel up!
Any small marker you leave there may be impossible to spot
again once you have flown away. How could you know
exactly where to fly back to?*

*Come back slowly to the room now, as I count 5 ...4 ...3
...2 ...1.... Welcome back! You can now discuss this problem
in your small groups. Record your suggested solutions, even
the 'wild' ones."*

Following this, indicate that the exploration they will do on these
pages might help later when you take up the jewel question again.
Suggest that in the meantime they should just get involved in each
part of the Activity for its own sake without bothering about how
it connects to this problem.

Answers

For each **Distortion Rule,** encourage students to write a "wild
guess" about what might happen, then give more than one simple
comment in their descriptions of how the ***Starter Picture*** has
changed. Conveying the exact changes in full sentences invites
clear, detailed, descriptive writing. Depending on how developed
their skills in writing and cooperative learning are, group drafts
and critiquing can be employed.

The ***Starter Picture*** should plot as an italic M

*Use expression, feeling, and
drama in your voice.*

*If students interupt with a noise
or giggle, simply give it little notice
and go on.*

*Allow a pause after they return
to the room. Begin speaking
slowly and quietly as you give
directions.*

jottings

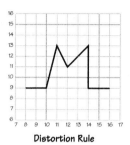

Distortion Rule

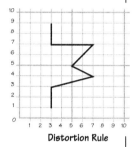

Distortion Rule

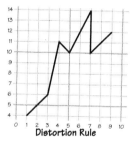

Distortion Rule

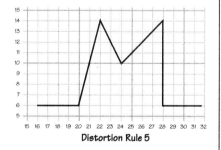

Distortion Rule 5

Distortion Rule 1: Adding 7 and 6 to the coordinates will cause the same picture to move to the right 7 steps and up 6 steps.

Distortion Rule 2: Reversing the coordinates will cause the same picture to spin 90° sideways, flip backwards, and land on top of the ***Starter Picture.*** *(See jottings at left)*

Distortion Rule 3: Adding the coordinates and putting them in the second place of each pair will result in a distortion of the M as shown at left:

Distortion Rule 4: Placing the sum in the first place of each pair turns the M of rule 3 sideways and flips it.

Distortion Rule 5: The size of the ***Starter Picture*** will be doubled and moved 14 to the right.

Distortion Rule 6: The ***Starter Picture*** will get 3 times as wide and 4 times as high when the first coordinate is tripled and the second is quadrupled.

Students are then asked to make up their own rules, and, in one extension, to make up their own group of pairs to make a ***Starter Picture***. Everything from the very simple transformations (like the first rule) to the more complex (like rules **3**, **4**, and **6**) can be devised and tried, depending on the abilities of the students. Even *combinations* of transformations (e.g., add something to one coordinate, multiply something times the other, then put the sum of the two results in the second place) can be tried. If students' imaginations wane, some leading questions like, "What about the product or difference of the coordinates?" or "What about first adding to, then multiplying, something times each coordinate?" might spark some exploration.

Negative Coordinates. For those who seem ready for more, encourage them to explore this more advanced activity with the full coordinate axes and **negative numbers**, using the special Graphing Sheet you have copied from the activity pages. If they are sophisticated and eager, they can be thrown into it to sink or swim and learn from their mistakes. After a few examples they will be at home with negative coordinates (outside of an occasional confusion of "across" and "up-down"). If they are more tentative, you can first give a side lesson on conceptualizing negative numbers with a thermometer or number line. They may also need some instruction on how to locate addresses with negative coordinates.

Their graph should make a picture of Math Mole. They will need to be very systematic with transformations of the coordinates, adding something to each first coordinate and subtracting something from each second, etc. Adding a negative number to another negative, -3 + -4, makes a *more negative* answer, -7. Adding some positive to a negative, -5 + 3, makes it less negative, -2. Trickier is to subtract a *positive* from a negative, -4 - 8, where the first dash is a *negative* sign and the second is a *minus* operation sign followed by a positive 8, making the -4 drop to a *more* negative answer, -12.

 Finding the Jewel. If you did the imagination exercise with the students involving the pilot trying to keep track of the location of the jewel in the flat savanna, it's time to ask them to reflect once more on a possible solution.

You can do the exercise now before proceeding.

The solution that is actually used for that problem by pilots is to obtain a bearing, or reading, of latitude and longitude by comparing two radio signals from known locations, one being Zinder's airport and another being, perhaps, the city of Maradi, not much further, but in another direction. Once the pilot's latitude and longitude are determined accurately, that position can be pinpointed easily on a return trip by standard electronic navigational methods (like the Global Positioning Satellite system).

The important idea here is that the whole earth can be thought of as a kind of flat grid on which every point has a numerical address. What are the axes and where do they cross? The horizontal axis is the equator itself. The vertical axis is a north-south line crossing through Greenwich, England and going all the way to the poles. If an address is on the west (left) side of that line, it is assigned a longitude number like "23° West" instead of -23 used on our negative number axes. The horizontal numbers can range from -180 (180° West) to +180 (180° East). Below the equator we would see numbers like "18° South" instead of -18. The vertical numbers can range from -90 (90° South) to +90 (90° North). The approximate address of our pilot's jewel in Niger is (15° East, 15° North).

Look at any flat map of the world to get the idea.

It is spoken of by navigators as "15° East, 15° North."

The important idea is simply that when students are finding points with respect to a set of axes they are doing essentially what navigators do to find their way around on the earth.

Naturalist Extension: Many animals, rocks, etc., look alike but one is a "distortion" of the other. Simple example: a mountain is a hill with its vertical (second) coordinates multiplied by, say 5. Have students draw a hill on the bottom line of 7-3, multiply some vertical (second) coordinates of its points by 5, and plot a mountain! Try it with other objects.

You can make this extension of the Cartesian coordinate concept as simple (using only positive axes, i.e., North and East coordinates) or complex (using negative, or South and West coordinates) as you like.

Africa is a huge continent — in fact, it's the second largest on earth, after Asia. It's divided into 53 different countries, and these include some island nations near its mainland shores. The populations of these nations range from 105 **thousand** to 151 **million**. You get to join Math Mole on a coloring expedition to a map of Africa. You'll color the map not just any old way, but according to some number thinking you'll do.

But first, to help you get to know your **map** (page 8-3) of all the nations of Africa, here's a **Map Quiz**.

a. Find the four **largest** countries (by area).
_____,_____,_____,
_____.

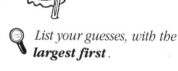
List your guesses, with the **largest first**.

b. Find the three countries that start with **T**: _____,
_____, _____.

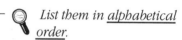
List them in <u>alphabetical order</u>.

c. Find the country with the most border made of **straight lines**: _____

Straight borders are man-made divisions done by Europeans in the late 1800's. Wavy borders are done by rivers or mountains.

 d. Find the country which has the most of its border touching the **ocean**: _____

e. Find the country that has the **most neighbors** touching it:
_____ Number of touching neighbors: ____

Map Coloring

f. Page 8-2 has the **"Populations of African Nations."** It shows all the African nations with their estimated 1997 populations. On your **map** of Africa, page 8-3, color each country a color according to **how large its population is**, using these **colors**:

 Red: Population of 50,000,000 or above

 Orange: Population of 20,000,000 to 49,999,999

Yellow: Population of 10,000,000 to 19,999,999

 Green: Population of 5,000,000 to 9,999,999

 Violet: Population of 1,000,000 to 4,999,999

Blue: Population of 500,000 to 999,999

 Brown: Population of 100,000 to 499,999

helpful   to do   calculate   write

2007 Populations of African Nations

in Alphabetical Order

Algeria	33,330,000	Libya	6,040,000
Angola	12,260,000	Madagascar	19,450,000
Benin	8,080,000	Malawi	13,600,000
Botswana	1,820,000	Mali	12,000,000
Burkina Faso	14,330,000	Mauritania	3,270,000
Burundi	8,400,000	Mauritius	1,250,000
Cameroon	18,060,000	Morocco	33,760,000
Cape Verde	424,000	Mozambique	20,910,000
Central Afr. Republic	4,370,000	Namibia	2,060,000
Chad	9,890,000	Niger	12,900,000
Comoros	711,000	Nigeria	135,030,000
Congo	3,800,000	Rwanda	9,910,000
Dem. Rep. of Congo	65,750,000	Sao Tome e Principe	200,000
Djibouti	496,000	Senegal	12,520,000
Egypt	80,340,000	Seychelles	82,000
Equatorial Guinea	551,000	Sierra Leone	6,150,000
Eritrea	4,910,000	Somalia	9,120,000
Ethiopia	76,500,000	South Africa	44,000,000
Gabon	1,460,000	Sudan	39,380,000
Gambia	1,690,000	Swaziland	1,130,000
Ghana	22,930,000	Tanzania	39,380,000
Guinea	9,950,000	Togo	5,700,000
Guinea-Bissau	1,470,000	Tunisia	10,280,000
Ivory Coast	18,010,000	Uganda	30,260,000
Kenya	36,910,000	Zambia	11,500,000
Lesotho	2,125,000	Zimbabwe	12,310,000
Liberia	3,200,000		

 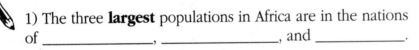

1) The three **largest** populations in Africa are in the nations of _____, _____, and _____.

2) The two **smallest** populations are in the nations of _____, and _____.

3) The population closest to **half a million** is in the country of _____.

4) The two countries with the **same** population are: _____, _____

helpful | to do | calculate | write

MAP of AFRICA

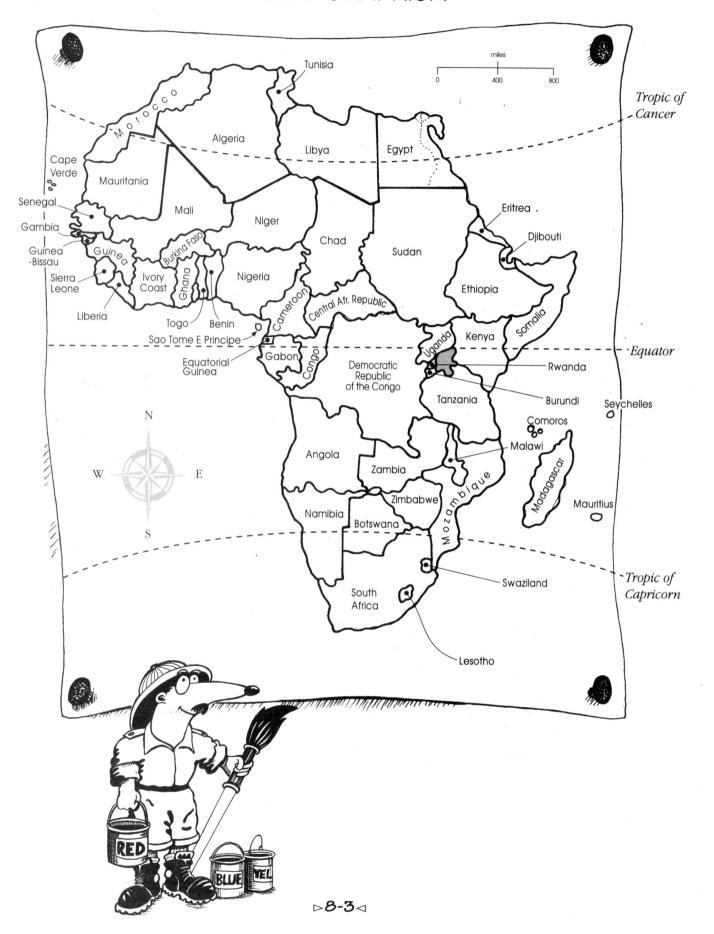

miles

0   400   800

Tropic of Cancer

Tunisia

Morocco

Algeria

Libya

Egypt

Cape Verde

Mauritania

Mali

Niger

Chad

Sudan

Eritrea

Djibouti

Senegal

Gambia

Guinea-Bissau

Guinea

Burkina Faso

Sierra Leone

Ivory Coast

Ghana

Nigeria

Cameroon

Central Afr. Republic

Ethiopia

Somalia

Liberia

Togo

Benin

Sao Tome E Principe

Equatorial Guinea

Gabon

Congo

Democratic Republic of the Congo

Uganda

Kenya

Rwanda

Burundi

Tanzania

Seychelles

Equator

Comoros

Malawi

Angola

Zambia

Zimbabwe

Mozambique

Madagascar

Mauritius

Namibia

Botswana

Swaziland

South Africa

Lesotho

Tropic of Capricorn

N   W   E   S

RED   BLUE   YEL

▷8-3◁

Country Graph

👉 **g.** Color a **rectangle** on the bar graph for each nation according ◄— to its population. Each column of the bar graph stands for a "**population range**."

For later use, also make a **colored mark** next to each country on the "Populations of African Countries" list as you graph it.

80,000-499,999	500,000-999,999	1,000,000-4,999,999	5,000,000-9,999,999	10,000,000-19,999,999	20,000,000-49,999,999	50,000,000-or more

✏️ **h.** Write at least five things the graph is telling you. Examples: "Most countries have populations greater than....," ◄— "There are ____ countries with populations greater than ____."

_____

_____

_____

_____

_____

To get started, look carefully at the graph. What numbers and relationships can you see?

The Mode

The category that has the most colored graph rectangles is called the **mode** of the data.

Several nations are in this range.

✏️ **i.** Using your **bar graph,** find the **mode** population range of the African nations. _____ Pick a **nation** that is in the mode population range on the graph to represent the mode of Africa. Write it here: _____

The Average (Mean)

If the people of Africa moved across borders until every nation had the **same population**, that population would be called the **average** (mathematicians call it the **mean**) population of Africa. You find this by lumping all the populations together (adding them to get a total) then breaking that total into 53 equal parts, one part for each nation.

j. Find the average (mean) population of all the African nations. Write it here: _____

Work with partners — split up the nations on the list into clumps, then **each partner** finds a **clump-total**. Since it's easy to make mistakes adding, **trade clumps** and total them again as a check. Finally add the checked clump-totals to create a **grand total** population of Africa.

k. Pick a nation with population closest to the average to represent the mean: _____

The Median

The **median** population is also a sort of middle-sized population. If all the nations' populations were arranged in a line from the **smallest to largest**, and then the **middle** number in the line was picked out, that would be called the **median**.

l. Work with a group of students to figure the best way to arrange the nations in order of their population size and pick the **middle** one. The **nation** with the **median** population is:

_____

What's "Typical" and "Medium?"

This thought activity compares the **mode**, the **mean**, and the **median** populations of Africa.

m. Could the **mode** nation's population be "medium" or "typical" in some way? How? _____
_____

n. Could the **mean** (average) nation's population be "medium" or "typical" in some way? How? _____
_____

o. Could the **median** nation's population be "medium" or "typical" in some way? How? _____
_____

p. Which one of the three above would you say should be the **best representative** or **best example** of the size of the African nations? Why? (Give a good argument.)_____
_____
_____

CALCULATOR TIP: Leave off the last three zeros of each population figure as you add. It will save time and your calculator screen will not overflow. After you get the grand total, divide by 53, and round to the nearest unit, then put three zeros back on to get the proper size for the average.

Population Density (more challenging)

The **population density** of a country is found by computing the ratio of its population to its area. That is,

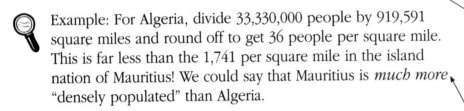

$$\text{POPULATION DENSITY} = \frac{\text{POPULATION OF THE COUNTRY}}{\text{AREA OF THE COUNTRY}} = \underline{\hspace{2cm}} \text{ PERSONS PER SQUARE MILE}$$

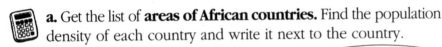

OR: DENSITY = POPULATION ÷ AREA

That is, we divide a nation's population by its area and get the average number of people in each square mile of the country.

Example: For Algeria, divide 33,330,000 people by 919,591 square miles and round off to get 36 people per square mile. This is far less than the 1,741 per square mile in the island nation of Mauritius! We could say that Mauritius is *much more* "densely populated" than Algeria.

If all the people were spread out evenly in the whole country (like jam on toast!) every square mile would have that number of people on it!

a. Get the list of **areas of African countries.** Find the population density of each country and write it next to the country.

For comparison, the U.S. has 31 persons per square mile.

b. 1) For **each nation** cut a small stick. It has to measure **one millimeter** for every **ten people per square mile.** (Divide the population density by **10** to get the millimeters.)
2) Stand the stick in a tiny blob of rubber cement or white paste on its country on the map.

For the stick, use one these materials: straw or plastic strands from a broom, thin plastic stirring straws, toothpicks, or spaghetti noodles.

For instance, Algeria's 36-density stick would be a tiny **3.6 mm** long (just a hair below 4 mm). Mauritius' 1,741 density would get a long 174.1 mm (or 17.4 cm) stick!

c. On another sheet, write some observations about population densities and their sizes in different parts of Africa. Comment on patterns of the sticks or connections you see.

First, discuss with others how the sticks make patterns across Africa. Try to explain why this is.

helpful | to do | calculate | write

AREAS OF AFRICAN COUNTRIES IN SQUARE MILES

Country	Area	Country	Area
Algeria	919,591	Libya	679,359
Angola	481,351	Madagascar	226,656
Benin	43,483	Malawi	45,745
Botswana	231,803	Mali	478,764
Burkina Faso	105,868	Mauritania	397,953
Burundi	10,745	Mauritius	718
Cameroon	183,568	Morocco	172,413
Cape Verde	1,556	Mozambique	309,494
Central Afr. Republic	240,533	Namibia	318,259
Chad	495,753	Niger	489,189
Comoros	719	Nigeria	356,668
Congo	132,046	Rwanda	10,170
Dem. Rep. of Congo	905,563	Sao Tome and Principe	371
Djibouti	8,494	Senegal	75,749
Egypt	386,660	Seychelles	176
Equatorial Guinea	10,830	Sierra Leone	27,699
Eritrea	46,842	Somalia	246,200
Ethiopia	435,186	South Africa	471,444
Gabon	103,347	Sudan	967,494
Gambia	4,363	Swaziland	6,703
Ghana	92,100	Tanzania	364,900
Guinea	94,927	Togo	21,927
Guinea-Bissau	13,946	Tunisia	63,170
Ivory Coast	124,502	Uganda	91,135
Kenya	224,961	Zambia	290,583
Lesotho	11,718	Zimbabwe	150,803
Liberia	43,000		

Use this for the Population Density activity.

helpful

to do

calculate

write

Teacher Notes

Intelligences emphasized: Visual-spatial, Interpersonal (cross-cultural), Kinesthetic, Linguistic, (Musical), Naturalist

Math skills, concepts: Comparing large numbers, calculator skill, area estimation, bar graphing, mean/median/mode, ratios (population density).

Materials: World globe, crayons or colored pens, paste or glue, scissors, and one of these (that can be cut into small measured lengths): straw or plastic strands from a broom, thin plastic stirring straws, toothpicks, or spaghetti noodles.

Level of difficulty: The estimation and coloring activities are moderately easy; the statistical activities are a bit harder; the population-density activity is more physically and conceptually challenging.

Overview: This activity creates a geography and social studies lesson in the process of teaching many math skills. (NOTE: South America, Europe, or Asia could also be used in this same activity mode.) It rotates around a map, a list of populations, and a list of areas.

The coloring will be very absorbing. Then students can cooperate in the detailed process of averaging and finding the median. Dividing up the tasks helps to avoid tedium and mistakes. Tips for this are included on the activity pages.

The work with population densities is very hands-on and visual-spatial, but it takes some understanding of the ratio idea: "Persons-per-square-mile" is called *density*. It tells how many people would occupy one square mile of a country if they were all spread evenly over the land."

Introduction to class: Hand out the first page of the Activity and the page with the map of Africa. A world map should also be available. One possible start of your presentation could be a brainstorm session in which students pool what they already know about Africa. Then bring out any or all of the kinds of information sketched below.

"Africa is not a country but a *continent*. Continents are large masses of land, usually separated by water from other masses. How many can you find on the map or globe? Africa is the second largest,

jottings

having 1/5 of all the land on the globe but only a bit more than 1/8 of the population of the world. Asia is the largest with almost 1/3 (actually 3/10) of all the land. What fraction of the world's land do you think North America has? [1/7]

"Much of Africa is near the equator, so it is mostly hot, but there are many different climates there. There are also many nations, each with its own government, products, landscapes, customs, and history. Each one of them was ruled, and often exploited ruthlessly, by a European nation in the past. All are self-ruling now and many are still trying to recover from this difficult past. [Social studies lessons, music experiences, artifacts, food tasting events, and travel films can augment this activity and can segue to other parts of your curriculum.]

"In Africa there are three more nations than there are states in the United States. Most of these nations are located on the continental land mass but a few are islands near it. Those are Madagascar (the huge island), Mauritius, Comoros, Seychelles, and Sao Tome.

"Some of these nations have high populations, some have very low ones. Why is that? What factors determine how much population a nation will have? [Size, usefulness of land, policies of government, climate, birth rate, death rate, access to ports, wars, etc.]

Answer the questions on the first sheet using the map. Then get pages 2 and 3 along with the list of "Populations of African Nations" list. Answer the questions on the bottom of the population sheet. You will start with a country-coloring and graph-coloring assignment as you follow the directions on the sheets."

Answers

Map Quiz
a. The four largest countries are, in order: Sudan, Algeria, Democratic Republic of the Congo, and Libya.
b. Togo, Tanzania, Tunisia
c. Egypt, Libya, and Algeria all have about the same length of straight border.
d. Madagascar, the very large island.
e. Democratic Republic of the Congo, 9

The **Populations of African Nations** Sheet Answers
1) The three largest populations are Nigeria, Egypt, and Ethiopia.
2) The two smallest populations are Seychelles and
   Sao Tome e Principe.
3) Djibouti is closest to a half million.
4) Sudan and Tanzania have the same population.

Map Coloring and Country Graph

The map and graph are colored the same way, essentially. They can be done simultaneously or separately. (If the student has made colored marks next to each country on the **Populations of African Countries** sheet while coloring the map, as advised, it *may* be easier to graph and it *will* be a lot easier to find the **median** later.) The graph, when finished correctly, should show all the countries' population categories as follows:

4	2	13	9	12	9	4
80,000-499,999	500,000-999,999	1,000,000-4,999,999	5,000,000-9,999,999	10,000,000-19,999,999	20,000,000-49,999,999	50,000,000 or above

h. Students answers will vary, and a class discussion is in order to discuss the graph's meanings. Some typical meaning statements (among many) are:

- "Most countries are between 1,000,000 and 5,000,000. but there are almost as many between 10,000,000 and 20,000,000."
- "There is a wide range of populations with only a few at the extremes."
- "There are 9 countries with populations between 5,000,000 and 9,999,999."
- "There are more countries with populations higher than the mode than there are countries with populations lower than the mode."

The Mode

The **mode** of the population categories is, from the graph, the population range 1,000,000 - 4,999,999. Any of thirteen different nations, Liberia, for example, could be chosen to represent the mode, which could be interpreted as the most "typical" size.

The Average (Mean)

j. The **average** or **mean** of the populations is 17,622,000.
k. The population of Ivory Coast is closest to it with Cameroon a close second. (If the averages in the class vary from this figure, have students go back and trouble-shoot their calculations by double-checking each group of ten numbers.) The mean is one candidate for "medium-sized" population.

jottings

The Median

l. The **median** of the populations of African countries is 9,910,000, the population of **Rwanda**, whose population shrank in the mid-90s due to genocide and refugee emigration. It has grown from about 8,000,000 then. This is the other candidate, along with Ivory Coast, for the "medium-sized" population.

What's "Typical" and "Medium?"

m. The mode is more a "typical" population than a "medium" population, because it appears the most. The mode here is not a very good "typical" value because the category 10-20 million is nearly the most common also.

n, o, p. The mean and median are both measures of a kind of "medium" value that represents the populations. Is Ivory Coast's 18.01 million or Rwanda's 9.91 million more "representative?" Both have their strengths and flaws as candidates.

When there is a single very large value in the data, i.e., Nigeria, at 135 million, with Egypt's 80 million a distant second, the **mean** is influenced upward by Nigeria, making it appear that the "medium" size of all the *other* countries is larger than it really is. (In fact, if Nigeria were left out and the other 52 averaged, their mean would be 15.4 million.)

The median, 9,910,000, doesn't care whether Nigeria is 151 million or 70 million, as long as it's still the biggest population. The median seems to register the fact that more than half are small populations, while less than half are very big populations. Of the two, mean and median, the median, Rwanda, would probably be the most representative here.

Population-Density. a, b. This activity is time-consuming but rewarding. It's an excellent group activity in which everyone's contribution of labor is welcomed. It forms a sort of three-dimensional graph. Trends will be noticed when students have finished.

c. For instance, there is a whole section of the continent where population-density tends to be lower. A good geography lesson is to look up these countries (in a fairly *modern* encyclopedia!) and determine why so few people would live in all that expanse.

The coastal nations tend to be more densely populated, with notable exceptions, like Sierra Leone. Looking up the nations that are exceptions to the trends is a valuable lesson in itself.

Sometimes students will find that certain sparsely populated countries have dictators, famines, a history of poor government, or poor land.

Extensions:
(Many research extensions are given in the **Answers** section.)

Average Population Density. Average the population densities of all the nations to get "the average population density" of the African nations. Which countries are near this average? Compare it to the U.S.

Research slavery. Mainly from which countries were slaves brought to Europe, starting with the Portuguese in the 1400's? To the New World starting in the 1600's? Why these countries?

African music. This rhythmic music was brought to the U.S. and Europe via slavery. African music was a major component and influence on the development of jazz and rock and roll in the U.S. Which African countries' music are we most indebted to? Find selections of music from these countries (Ghana and Nigeria being easiest to find, then Senegal, Guinea, Gambia, and Mali) or from groups claiming to play "African rhythms." Play them, listening for connections to our own music. Find a few selections of music, if possible, from countries where U.S. slaves didn't originate (e.g., South Africa, Egypt, and Madagascar). Does it sound more foreign?

Demographics. Do you think population numbers of slave countries are different than they would be had there not been the removal or death of twenty to thirty million inhabitants caused by the European and American slave trade in three centuries? [A hard question to answer, there are so many factors that influence population size in three centuries.]

Map Work. The area of the United States is 9,666,861 square miles. Find a connected group of African countries that are about the size of the United States. Outline these with a colored pen on the Africa map.

Species Populations. As a project for the naturalists, research a handful of animal or plant species found in parts of Africa. Part of the report is a coloring of the map to show where the species are (a different color for each species). It can grow more sophisticated if students speculate on why the species are found where they are.

jottings

PATHS AND PLANETS

Planets move according to certain rules. The **rules** make the **shape** of their orbits. You'll see how rules make shapes if you work with your cm **ruler** — a **compass** is handy too. You'll work with planets later, but first it's best to start closer to home with a **goat** making a path.

 a. Staked goat. At the right is a stake driven into the ground. Suppose your goat **Daisy** is tied to that stake with a rope so that her mouth can't get any farther than the rope length from the stake. Draw a small, simple goat facing away from the stake, with the taut rope tied to its collar.

Daisy loves to strain against the rope and nibble all the grass she can find. Using your **compass** carefully draw the shape of the grass patch Daisy will be able to "mow."

 b. Leashed goat. Now it gets a little harder. At the right are **two** posts with a wire tight between them.

Daisy's leash is attached to the wire it so it can slide freely along the wire. Draw a small Daisy pulling at the end of the leash.

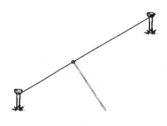

Use your ruler to measure and draw the shape of the biggest grass patch Daisy will be able to "mow."

 c. Jealous kids. This time Daisy has another problem. She has two kids, **Rip** and **Bif**, that are **very** jealous of each other. They're having to wait in tiny **pens** pictured at the right. Daisy would like to take a walk, but she has to be very careful. **Whenever** she is the **slightest** bit closer to Rip, **Bif** goes crazy. If she is just **slightly** closer to Bif, **Rip** goes crazy. Draw a long path she could walk where **neither** will go crazy.

 Geometry students call the shapes you drew in
a. "the set of all points that are the same distance from one point;"
b. "the set of all points that are the same distance from a line segment;"
c. "the set of all points that are the same distance from two points."

Rip   Bif

helpful

to do

calculate

write

d. Round and long cages. Now Daisy is in a very tricky situation at the right. Rip can only walk back and forth inside the long narrow cage. Bif is in the small pen. Rip **always** walks in the long cage to the place where he is **closest** to Daisy, no matter where she is.

Rip or Bif goes crazy any time Daisy is closer to one than the other. Daisy is eager to take a walk. **What whole path can she follow**, in both the **right** and **left** part of the space shown, so that Rip and Bif will be happy? (Two places she can be are shown.) Make some dots, then draw it.

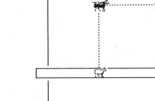

The path she wears into the grass as she walks like this every day is called a **parabola** (say *pear-a-bowl-uh*, where the *a* sounds like the *a* in *a*pple). This is a mathematical shape that is **not** a piece of a circle.

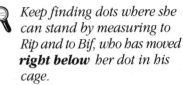

*Keep finding dots where she can stand by measuring to Rip and to Bif, who has moved **right below** her dot in his cage.*

*The **parabola** curve shows up in **three** common places in our lives: throwing a ball in the air; sattelite TV dishes; lights, like flashlights, car lights, and spotlights. Can you tell how?*

e. The goat and the ellipse path (more challenging). The last situation Daisy needs to deal with is this. **Two stakes** are in the ground, but there's a loose rope tied between them. Daisy has a loop on her collar that can slide anywhere along that rope. She always likes to pull away from the stakes as far as she can go and keep the rope taut, eating the grass she can reach. Outline the patch of grass she can "mow."

Hint 1: Make several dots whose distances from the two stakes **add up to 7 cm.** For example, keep measuring 6 cm from one stake while sliding your ruler so its end is 1 cm from the other stake. Or make a dot that is 4 cm from one stake and 3 cm from the other, etc. Then fill between the dots.

Hint 2: If this is too difficult, go to the next activity "A String Drawing," do it first, then you'll really understand this activity. Return here later.

String-Drawing the Ellipse-Path

a. Find a larger-than-usual piece of **blank paper** and make **two** dark dots on it **30 cm** apart (see the figure).

b. Take a piece of string about **45 cm long** and tape one end to each dot.

c. Hook your colored marker point on the string. Pull gently against the string and begin moving it along on the paper. You will be forced to draw a curve (see the figure).

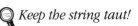

Keep the string taut!

helpful to do calculate write

d. Continue the curve until it completely surrounds the two dots.

 This figure you drew is called an ELLIPSE and the two dots are called its FOCUS POINTS.

 It's OK to lift your pen and untangle the string as you go.

*Notice that this string drawing is exactly the same as the Daisy goat problem **e**.*

Ellipse Experiments

1. Use the **same (45 cm) string** for all of these. You will use **different focus points**:

 a. Make <u>another ellipse</u> the same way, but with the two focus points <u>farther apart</u> than 30 cm.

 b. Make <u>another ellipse</u> the same way, but with the two focus points <u>closer together</u> than 30 cm.

 c. Make both focus points stand together as <u>one point</u>. Attach <u>both</u> string ends to that point. Use the same ellipse-drawing method.

 d. Write, in complete sentences, your conclusions about how **ellipses** change when **focus points** change:

_____

_____

 Save all your ellipses for activities you will do later!

2. Use the **same (30 cm) focus points** but **different strings**:

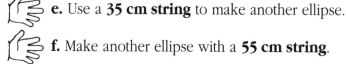

 e. Use a **35 cm string** to make another ellipse.

f. Make another ellipse with a **55 cm string**.

 g. Make a third ellipse with a **70 cm string**

 h. Write your conclusions about how **string length** changes an ellipse:

_____

_____

 You'll need bigger paper for some of these.

Save these ellipses for later too.

Describing Ellipses — A Cooperative Game

You now have several ellipses. You've seen that they can be nearly a circle, or long and narrow. Here's a detective game to try in a group:

 a. Each member of your group should pick one of the ellipses made earlier **without showing it to anyone else**.

 helpful   to do   calculate   write

b. Each player should use a ruler secretly on one of his or her ellipses to take **three** straight measurements that could tell the group about the ellipse's shape and size. Label what each number means on the ellipse.

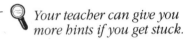

Measurements:
1. _____, __.___ cm
2. _____, __.___ cm
3. _____, __.___ cm

 c. Take turns telling the group the measurements taken and where they are on the ellipse. Everyone should try to guess what the hidden ellipse looks like. Is it long and flat, nearly circular, large, small?

The amount of "stretchedness" an ellipse has is called its "eccentricity."

(Challenging) A Flatness or Longness Number. Try to calculate from any of your measurements a **single number** that tells how much "stretch" or lengthening your ellipse has compared to a circle. Try to make your number calculation stay the same for any ellipse, whether it's tiny or huge, that looks as stretched as yours.

Hint: Think of ratios of your measurements. They don't depend on the size of the ellipse — only its shape.

Your teacher can give you more hints if you get stuck.

Ellipses in Outer Space

The great German astronomer **Johannes Kepler** wrote in 1609 that he had discovered this:

> All planets, in fact **all** objects in space that orbit around other objects, don't make a circular path.
> They trace an **ellipse**.

The other amazing thing he discovered is:

> Whatever body the objects orbit around is located right at **one focus point** of the ellipse.

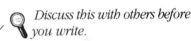

Things that orbit around other bigger things in space:

rocks   comets
moons   planets
stars   galaxies

 a. Do you think a comet following the path of your ellipse around Saturn always travels the same speed? _____
Why? _____
_____

Earth and **Pluto** follow elliptical paths around the sun, but both ellipses aren't the same shape. Earth's 186,000,000 mile diameter orbit looks **just like a circle** at the right. It's two focus points, 3,000,000 miles apart, look like **one dot**!

Pluto has the most "eccentric" or elongated orbit. It is **9 billion** miles in length and the distance between its focus points is 2 1/4 billion miles! It looks about like the picture. The Sun is at point S.

Discuss this with others before you write.

Earth's Orbit

Sun

Pluto's Orbit

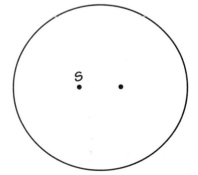

helpful   to do   calculate   write

b. Give measurements that prove Pluto's orbit is <u>not</u> a circle.

_____

 c. Try to draw **earth's orbit** the **right size** inside <u>Pluto's</u> orbit. ◄———— _Think and compute carefully!_

Ellipse Travel •———————

Below is a very stretched or "eccentric" ellipse in outer space. It is like the shape of the path of a comet. Many comets buzz around giant Jupiter, looking tiny here at one focus point of the comet's path — a path that's long because the focus points are **far** apart. It will take the comet a long time to go around Jupiter once.

**Comets** are everywhere inside and outside our solar system!

On the comet's ellipse, think of each number as a speed.
13 is the **fastest** speed it goes while near Jupiter.
7 is a **medium** speed and
1 is the **slowest** speed the comet goes at the far end.

Jupiter · 13 · 1 · 7 · 3 (ellipse diagram)

 A comet doesn't go the same speed around its whole ellipse. Going **toward** Jupiter's pull it **speeds up**, then "slings" around it at the closest approach. That starts a "climb" away from the pull, **slowing it down** until it's at a "snail's pace" at the farthest end of the loop, where it turns and starts all over.

 The time the comet travels **1** space at the far end is the **same time** it travels **7** spaces when past the middle and **13** spaces near Jupiter!

 a. Try to move your finger around the orbit so you keep the right speed at each place. You have to practice several orbits to get smooth speeding up and slowing down all the way around.

◄——— _**Hint:** Make your **1**-speed **very slow**._

 b. Have a partner check your comet-finger's movement for accuracy. Then you should check your partner, until both of you are sure you have the right movements and speeds all around the ellipse.

 helpful to do calculate write

Teacher Notes

jottings

Intelligences Emphasized: Spatial, Bodily-Kinesthetic, Interpersonal, Linguistic

Math Skills, Concepts: Geometric relationships (locus of points), metric measurement, estimation, large numbers, parabola, ellipse

Materials: Paper, string, tape, large printed or unprinted paper, centimeter ruler, calculator

Overview: This Activity informally studies the notion of a locus, which is a shape determined by rules or parameters. It then segues from this concept to a figure well known in classical mathematics and in astronomy, the ellipse. Not just an oval or egg shape, an ellipse has very precise criteria, so it is a locus. These include two focus points (kind of like what would happen if the center of a circle split into *two* centers that continued spreading apart, causing the circle to become elongated).

The ellipse has a special place in the heavens — it's the shape every orbiting body takes as a path around a bigger body. Students have a chance to get their hands involved making ellipses, studying what influences their shape, and developing ways to measure and describe them.

Introduction to Class: It's best to organize the students in small groups or pairs. Floor or table space is helpful for drawing the ellipses. While getting organized for the activity, or while they are working, you could play the Pluto part of the classical selection of "The Planets" by Holst.

"In this Activity you will start off with the shapes goats make when they are restrained in certain ways. Each shape could be thought of as a set of points that keep a certain rule. One of the shapes that the goat, and you too, will make is the **ellipse** You may have heard of it — what do you think it is, and where do you think it occurs in the world or universe?"

To answer the question, refer to the fact that orbiting bodies travel along elliptical paths and that they will learn how in this Activity.

Hand out only the first couple of pages to start.

jottings

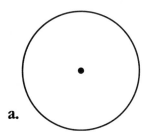

a.

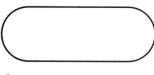

b.

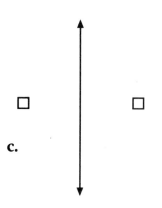

c.

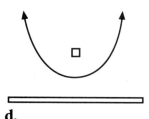

d.

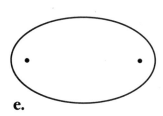

e.

Answers

The Goat Problems

The parameters in **d** create a figure called a **parabola,** and Bif's small circular cage is called its **focus.** It's the set of all points that are equidistant from **both** the **focus** and the **line** (the long cage). (Distance from the line must always be "straight to," that is, perpendicular to, the line.)

The **parabola** goes upward forever. The cut-off piece drawn for the goat, turned upside down, and stood on the ground, is the mathematical path gravity causes a thrown ball to take when it rises into the air and falls back to earth. A ball thrown more steeply creates a narrow parabola that still looks like the goat path if it were extended much longer and seen from far away. A ball thrown hard almost horizontally still arcs up and down, and its path is like a tiny slice from the bottom part of the goat curve.

The **parabola's** shape has a reflective characteristic that causes all light generated by a bulb at its focus to reflect from its sides straight out its mouth, or all radio waves coming straight into its mouth to converge at its focus. (Pretend the parabola has a shiny surface and Bif's cage is a light bulb. Draw ray-lines leaving the light bulb, hitting the parabola, reflecting, and going straight out the mouth of the parabola.) Thus it is also used as a reflector shape in headlights and in microwave antennas. A separate presentation on the parabola, or its exploration in a math/science student project, is very absorbing to curious students; parabolas can be researched on the world wide web.

e. The answer is an ellipse because every point of an ellipse has the same sum-of-distances to the two focus points. This serves as the very definition of an ellipse.

String-Drawing the Ellipse-Path

Drawing the ellipse takes careful bodily-kinesthetic focus. Here are some hints for better drawing:
- The string is supposed to start out slack with the ends anchored at the two points.
- Maintain tautness while drawing, without pulling *too* hard on the taped string.
- Keep the string just hooked on the pen point without sliding off.
- Keep the pen upright at the same angle as you move along.
- At the extreme ends of the ellipse the string will be looped around the pen and needs to be repositioned to go on.

Point out that the string is the distance-sum to the focus points and that it remains the same length for each ellipse point. The goat problem ☞ **e** above it presents the same criterion.

Ellipse Experiments

 1. d. As the focus points move closer together the ellipse becomes more circular. (When they coincide they make an exact circle.) When the points move farther apart, the ellipse gets more elongated.

2. h. As the string grows longer the ellipse grows more circular, and when it is shortened the ellipse gets more elongated.

Describing Ellipses — A Cooperative Game

b. The three distinct measurable lengths on an ellipse are:
- Its **length** along a center line through the focus points (foci),
- Its **width** or height across its center directly between the two foci, and
- The **distance between focus points**.

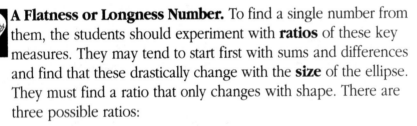 **A Flatness or Longness Number.** To find a single number from them, the students should experiment with **ratios** of these key measures. They may tend to start first with sums and differences and find that these drastically change with the **size** of the ellipse. They must find a ratio that only changes with shape. There are three possible ratios:

- The length-to-width ratio does this, changing from 1 for a circle to 2, 3, 10 and beyond as the ellipse becomes cigar-shaped.

- The focal-length-to-width ratio does this, changing from 0 for a circle to 1, 2, 5, 10 and beyond as the ellipse elongates.

- The focal-length-to-whole-length ratio goes from 0 for a circle to .9 for a very stretched ellipse but never can reach 1 no matter how long the ellipse becomes.

- Astronomers have chosen the latter and called it the **"eccentricity"** ratio. For earth's orbit it's .017, and for Pluto's more stretched orbit it's .25. This means that Pluto's focus points are a distance apart equal to 1/4 the length of the ellipse. Those more motivated students who simply explore these ratios and their behaviors will come to a sophisticated appreciation of ellipses.

The earth's orbit looks totally circular in any illustration but it truly *is* an ellipse that is 3,107,000 miles longer than it is wide. Only Venus and Neptune have even more circular orbits.

 b. The student need only measure the length and width of the drawing of Pluto's orbit to see that they are different.

 c. The ratio of the Earth's ellipse length to Pluto's ellipse length is .0206. This works out to earth's *whole orbit* being about 1 mm, or only the big dot, marked S in the illustration!

Ellipse Travel

 a, b. This is a kinesthetic experience of the travel of a comet around a gravity point. Students can report orally or in writing about their thoughts and experiences afterwards. They may find it tricky to find the right speeds. A timer is helpful so that they can time five seconds to go 1 unit when away from Jupiter and the same five seconds to travel 7 units then 13 units as they approach Jupiter.

Extensions:

Research the planets. Find other information about their orbits, like the way they are tilted with respect to their planes of revolution about the sun.

Ellipse Race. Make an ellipse in a large area, like a playground or park. While two people hold the string at the foci that are about 25 feet apart, a third walks and marks off a large ellipse while holding taut a long (35 foot or more) piece of string. Others use markers, chalk, rocks, or markings in dirt to make the ellipse show up. Mark the two foci well with rocks or objects.

Two students compete by standing at each focus. Two others are "markers" on the boundary of the ellipse, and each "marker" is paired with one of the focus students. At the sound of Go! both focus students must race to their marker students then turn and run to the opposite focus postion. The first to make it wins. In the process they will realize again that the position of each marker student is of no real advantage because the distance to be run by each student is always the same.

Comet race. Two students play comets that start at each end of the large ellipse described above. One student plays Jupiter at one focus. Both comets start around the same direction and go their appropriate speeds for their positions with respect to Jupiter. Jupiter "disciplines" the comets if they are going too slow or fast in their orbit. Neither comet should catch the other.

Auditory ellipse. Students sit at a focus of the large ellipse with their eyes closed. A student or teacher rings a small bell or makes a tone on an instrument as he/she walks around the ellipse at a constant speed. Students describe their auditory impressions of the shape of an ellipse. Then the sound-maker walks around the ellipse at speeds corresponding to a comet's movement around Jupiter and the listener gives impressions afterwards.

Calligraphy. Many methods of calligraphy, especially those championed by the acclaimed expert Lloyd Reynolds, emphasize that all curves on the letters should be parts of ellipses, and that many correctly executed letters can just fit into an ellipse. In other words, the ellipse is a constant reference in the mind of the calligrapher. Introduce calligraphy to the students as a segue into art and language from this Activity.

Make an ellipse with lines (see the jottings).

1. Draw a circle of about 10 cm radius with a compass.

2. Mark its center point and another point F toward, but not at, the edge of the circle (perhaps 2-4 cm from the edge).

3. Take a piece of thin cardboard with a square corner and mark the long left edge with a red colored pen (that will show on the front and back sides of the cardboard). Mark the bottom edge with blue that can be seen on both sides.

4. Now, in the circle, touch the blue edge of the cardboard against point F and make the corner **N** touch the circle, at the same time. Hold fast and draw a line along the cardboard's red edge inside the circle.

5. Draw several such lines, *always* having the blue edge against F and the corner N touching the circle.

6. Make N touch the circle at many different points all the way around it and draw a line each time along the red edge. (To get all the way around the circle you will have to flip your cardboard over half-way through.)

If this is done right, an ellipse results with edges made up of many straight lines and with F as a focus. Mark another point for the other focus.

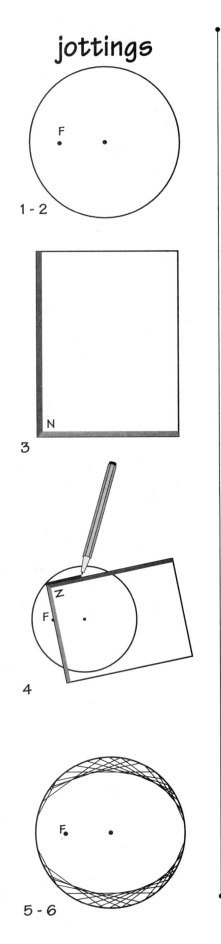

jottings

1 - 2

3

4

5 - 6

Data means information — often in the form of numbers. If you organize data right and think about it like a detective, you can predict the future! You'll see how.

How do you organize data? Sometimes it's **graphed** or sorted out in a **table**, and sometimes it's turned into numbers called **statistics**. If 75 National Basketball Association players are measured, that's *data*. When their **average height** is calculated, that's a *statistic*.

Now suppose these 75 players were a **randomly** chosen sample from all the hundreds of NBA players. Number detectives, called **statisticians** (pronounced *sta-tis-tish-uns*), call this a **sample average.**

"Randomly chosen" means "by chance," like drawing their names from a jar of the hundreds of names of all NBA players.

They know the **sample average** predicts the average you'd get if you measured **all** NBA basketball players. So it saves them a lot of time if they wanted to know the average of the whole group!

Now you can practice the tricks used by number detectives. Here are some cases for you to solve:

Case I: WORDS

Predicting from counts and graphs

 a. Cut an article from a newspaper or magazine. Pick any **one-foot-long** chunk of the article by drawing a line at its top and bottom.

 b. Count the 1-letter words, 2-letter words, 3-letter words, and so on, in the article. List your counts.

*Use **colored marker** pens to draw through the words as you count them. Color 2-letter words differently than 3-letter words, etc.*

 c. Make a **bar-graph** of your counts on another sheet.

Ask your teacher how to bar graph if you are uncertain.

For instance, if you marked 22 3-letter words with orange, you would color the 3-bar of your graph orange up to the number 22.

 d. Which word-length occurred *most* often? _____

This is called the "mode" of your graph.

 e. Which word-length occurred *least* often? _____

helpful | to do | calculate | write

f. Predict what the most and least-used word-length might be in another article. _____ Check **another** 1-foot piece of an article. What is its most common word length? _____

Predict from the data you have. But is your article a good sample of another article or the whole English language? Discuss your results.

g. Predict what would be found to be the most common word-length in the English language. _____

h. Give at least two reasons why your experiment might _not_ accurately tell you what the most common word-length in another article or the whole English language is:

Discuss your reasons with others, if possible, to get better ideas before you write.

_____

_____

_____

Predicting from fractions

i. Figure out how many words were there altogether in the article _____

j. Find these **statistics** — the _fraction_ of **all** the words in the article that have:

3 letters. _____	8 letters. _____
4 letters. _____	9 letters. _____
5 letters. _____	10 letters. _____
6 letters. _____	11 letters. _____
7 letters. _____	More than 11 letters. _____

Make a fraction by putting the total words (**i**) on the bottom (denominator).

k. Change each fraction above to a **decimal** and write it after the fraction, if you are comfortable with decimals.

Make decimals by dividing the top of the fraction (numerator) **by** the bottom (denominator).

Suppose you were blindfolded and stuck a pin into the article so that it hit a word. Do a **prediction** by giving the probability (fraction or decimal) that you would _stick_ a **6-letter word**. _____

Both probabilities are very related to the fraction or decimal you got in **j** for 6 letters.

What is the probability that you would _not stick_ a 6-letter word?

m. Keep sticking a pin into your article while you are blindfolded. Every time your pin sticks a word, record its size. Stick and record **40 words** this way. What fraction (decimal part) of the words that your pin sticks are 6-letter words? _____

Working with a fellow student and a tally-sheet for results makes this easier.

n. Try 40 word-pokes with a pin, as in **m**, on another article and see what fraction (decimal) of the words you stick are 6-letter words. _____

helpful   to do   calculate   write

o. Are the probability in **l** and the results in **m** and **n** about the same size? _____ Give reasons why you think they are similar or far apart.

_____

_____

*Discuss your results from **m** and **n** in a group to find your reasons.*

Averaging. What is the *average* word length in your article? _____

Multiply each word-length by the number of words of that length, then add up these products. Divide by the total number of words in the article.

Change something. Make up and try another experiment like the ones in **m** and **n**.

What a sentence! Just for fun, try to make up a sentence that has every word length from 1 to 11 in it.

_____

_____

It's even harder to use only <u>one</u> of every word length from 1 to 11.

Be a scientist. Make up another experiment of any kind involving word lengths. Make a plan, do your experiment, then write up your results. In your write-up, **describe** the experiment first, then make a **chart** or **table** of results, and include detailed **written descriptions** of what your numbers mean so that anyone could set up and check your experiment.

This is best done working with others.

Case II: COLORED COUNTERS

For this Case you need 66 colored counters — 36 of one color (let's say blue) and 30 of another (like red).

These can be colored cubes, squares cut from colored paper, poker chips, or tokens.

a. Spread out the 36 blue counters on one part of a table or desk in a close-packed cluster. Arrange 30 red ones the same way about a foot away.

One person volunteers to be *blindfolded*, and is called the Mover. The Mover must do the **"MOVER ACTIONS"** listed in **c** below, but first, just read the **"MOVER ACTIONS"** in **c** carefully.

Others can help direct the Mover's hand toward the middle of the right pile.

b. Make a bet. Predict, before the Mover does anything, what **chance** there is that this will happen:
After the **MOVER ACTIONS** *are done, the number of red counters in the blue cluster will be the* **same** *as the number of blues in the red cluster.* _____

Prediction: *Chance ____ Use a number somewhere between 0% (it will never happen) to 100% (it is certain to happen).*

helpful   to do   calculate   write

c. Do the **MOVER ACTIONS**:
- First **say** a number between 7 and 20; the example here will use "9."
- **Separate** 9 counters from the blue cluster and **mix** them into the red cluster.
- **Choose any** 9 counters in the mixed cluster, and move these over to the to the *blue* cluster, mixing them in.

The mover stays blindfolded, for all of these actions.

*Make sure the Mover returns the **same number** of tokens to the blue pile.*

d. Do the experiment at least five more times. Keep track of how many blues are in the red cluster and reds in the blue cluster each time.

You're a scientist doing experiments!

e. Try <u>changing</u> some things, like the number of starting tokens in both piles, the Mover-number, or even making the Mover-number different when moving back to the blue pile. **Keep a record of your changes and results!**

Change your prediction if you want.

f. Write a short paragraph that explains the results you got in the experiment.

This will take some discussion. Come up with a group explanation and write it carefully.

Mixing pop. A well-known problem goes something like this: "A can of root beer and a can of orange soda are each opened. A teaspoon of the root beer is mixed into the orange soda. Then a teaspoon from the orange soda is mixed into the root beer.

Question: When the mixing is finished, is there more root beer in the orange soda or more orange soda in the root beer?"

Get two empty pop cans to look at as you think about this problem.

You know the answer after working with the counters!

Try this problem on a smart adult you know. The adult can't just answer with a guess, but must include a **correct explanation**.

The probability is great that the adult will struggle with this problem and give a very complicated answer (that may be wrong)! The adult may not even believe your answer until you show the counter experiments you did in Case II.

Case III: DICE

You need three regular dice. You will gather and analyze data from dice throws. You will be able to *know* whether some dice games are fair or not.

Your group will throw all three dice 80 times while keeping a tally of the **product** of the three numbers. You need only mark an **O** for an odd answer or an **E** for even.

*Remember: <u>product</u> means **multiply**!*

a. Predict first how many of the 80 throws you *think* will have an **odd** product and how many will have an <u>even</u> product: Odd_____; Even_____.

Just make a good guess.

helpful — to do — calculate — write

b. Go ahead with the dice throwing, multiplying, and tallying 80 times. **Give your results here:** Odd_____; Even_____.

No calculator! There's a shortcut to figuring out whether the product will be even or odd. Can you find it?

c. Did your results agree with your prediction? _____
Write a clear **report** of your results and your thinking about why they came out the way they did.

Use another sheet and try **sketches, charts, tables, or lists** to make your report clearer. **Discuss** ideas with others before you write.

Make it Fair. Pretend there's a game where one person is "Odd" and the other "Even," and they take turns throwing three dice as above. Even almost always gets more points and wins. Design a new game of "Three-Dice Throw" so that Even and Odd have an equal chance of winning.

Add and Multiply. Check this game for fairness. For a play, each player, "Even" or "Odd," throws the three dice **one at a time.** The first two numbers thrown are **added**, then the answer is **multiplied** by the last number. Will Even or Odd win? **Predict**, then **play** 15 throws each, then **report**, explaining how the game is fair or unfair. If it's unfair, re-design the game to make it fair.

Case IV: VEHICLES

In this Case you will gather data on vehicles around you, then make some predictions. You will take a poll of vehicles that go by as you watch from a safe place by a road.

This is best done with other students so that more vehicles can be surveyed.

You and your partners should have copies of the VEHICLE TALLY SHEET, and each partner needs to watch a different place and enter descriptions of the vehicles that go by.

Make sure you're close enough to read the brand name of each moving car.

a. Gather data on at least **100 vehicles**.

b. Make a **separate** bar graph **for each of the five columns** of the VEHICLE TALLY SHEETS. For example, one bar graph would show the vehicle brands for all the vehicles.

For example: The bar graph for the "Make or Brand of Vehicle" column should have bars whose lengths show how many of each brand name you saw.

c. You can also compute other statistics. Find the average number of people in a vehicle: _____

d. The average number of people in SUVs: _____
in trucks: _____
in cars: _____

Find the averages in **c**, **d**, and **e** to one decimal place which means a tenth of a person!

helpful | to do | calculate | write

e. The **fraction** of all vehicles with only one rider: _____

f. Compute another **two statistics** from your data: ←

Use the counts or numbers in your data to find more fractions or averages.

_____

_____

g. The most popular color for vans: _____

for trucks: _____

for cars: _____

The most popular brand of van: _____

of truck: _____

of car: _____

These statistics involve only counting, not computing.

h. Find other counts from your data that are significant: ←

_____

_____

If a <u>group</u> found the data, make a group prediction. Compare your score with other group or individual scores.

i. Predictions: Using your data and statistics, make some predictions about the *first* moving vehicle you would see if you went back to that street. Predict the most likely make, color, type, etc. you would see:

_____

_____

_____

j. Go back to that road. Score **five points** for every characteristic you predicted correctly for the first vehicle you see. Score **one point** for each prediction that's right for the first five cars you see. Score: _____ Why do you think your scores were high or low? _____

_____

Discuss why your scores were high or low, with others.

People Survey. Select about five or six different characteristics (like hair color, race, glasses, shoes, sweater, etc.) and do a survey of people who walk along a street. Make several statistical statements about the people, similar to the ones that you made about cars.

Sit in an inconspicuous place with a good view of the walkers.

Plant Survey. Do a survey similar to the People Survey above, but with plants. Pick out a small area that has about 200 plants of any size on it and make a boundary with string. Make a chart like the vehicle one that names about 6 or more qualities that the plants in the small area have. Tally these in your whole area and summarize your data with statements like those in **c-h** above.

helpful | to do | calculate | write

VEHICLE TALLY SHEET

Use codes like R for red, Bk for black, Br for brown, C for car, V for van, etc. Show other codes you are using here:

Make or brand of vehicle (Ford, etc)	Main color of vehicle	Type of vehicle: (SUV, truck, car bus, other)	Number of riders Male	Female	Driver: male (M) or female (F)

Teacher Notes

Intelligences Emphasized: Kinesthetic, spatial, linguistic, interpersonal, Naturalist

Math Skills: Tallying, graphing, probability, mental multiplication, communicating results, hypothesizing, finding and comparing fractions and decimals, averaging, working with statistics.

Level of Difficulty: Challenging, (though less experienced students can learn from the first part of each sub-activity, omitting the more analytic questions).

Materials Needed: Paper, pens (of a dozen colors), colored counters (or scissors and colored paper), dice pencil, ruler, calculator

Overview: This activity will give students experience in actively gathering data, graphing it, deriving statistics from it, predicting from those statistics, as well as interpreting and explaining them. Four different arenas, or "Cases," have been chosen for work by number detectives.

Any one of the Cases, or even part of one, can be selected as a classroom activity for a given day. There is much rich discussion and interpretation that can evolve from each lettered request. After an introduction to a Case, some of the data gathering can also be assigned as homework. Data can then be pooled with that of others in a group and processed in class.

Introduction to Class: Ask students for their ideas of what they think a statistic is. Have them consult the dictionary as well as their own background knowledge. The dictionary will say something like "numerical facts or knowledge, assembled, classified, or tabulated so as to present significant information about a given subject." The introductory paragraphs of the Activity also make this distinction between *data* and *statistics*.

Discuss the role that statistics play in our society. An example is in the first paragraphs of the Activity, but there are many others — insurance, elections, the census, marketing. In none of these arenas can people deal individually with all the data available, so they have to "assemble, classify and tabulate" mountains of it first in order to see patterns in the data and to reason with those patterns. This kind of interpretive activity is very similar to what police detectives and plane-crash investigators do.

From their reasoning, detectives and statisticians can, in a way, predict the future. They do it with probability or predicted numbers. A fire investigator can state the probability that an arsonist will strike in a certain area, or an insurance statistician (actuary) can predict how many people in a group will die by age 75.

Indicate that students will get a taste of being number detectives or statisticians as they actively work on Cases in cooperative groups. Warn them that some requests in the Cases will require time and patient thought. Encourage them to work systematically through the alphabet letters of tasks, and parcel out jobs — e.g., counting, tabulating, graphing, writing and checking — so that work is done efficiently. To repeat, one, or even part of one, of the four "Cases" (which can be chosen in any order) is enough for a work period. More tasks or light-bulb extensions can be assigned for homework sessions.

Answers

Case I: a-h. The students will need to use a tally sheet, then a bar graph. Both will require some discussion about what form to use. The use of **color** is key, e.g., if orange is used to mark 3-letter words, it is also used to mark the tally of these, and it is used to color the 3-bar of the bar graph to a height that indicates the *number* of 3-letter-words.

Along the bottom of the bar graph students will need the numbers 1 through about 14, representing word-lengths. The side scale, number of words, should go up to about 28. Each side division can represent 1, 2, 3 or 4, words, depending on how compact the graph is to be.

The idea behind questions **a-h** is that the word-length that occurs most often in *this* article becomes a predictor of the most used word-length in *another* article. For question **f**, students may be able to do an abbreviated process to find the word-length that is most common. If the second article chosen for **f** is from the same publication, the results may be similar, but if it is from a *different* publication, the results may differ, and this difference should be discussed.

The article also becomes a predictor of the most common word-length in the English language. However, **f** and **h** raise the question that it may not be a very accurate predictor. This is true for many reasons; for instance, the article chosen may not be addressed to adults and thus may not use many big words. Even most newspaper articles tend to be written only to an eighth grade vocabulary level, which could mean that there are a lot more big words not used in newspapers. On the other hand, there are *many* smaller-length words, like *dearth*, that are still adult words.

Dictionary Extension: A faster group could survey word-lengths on about five random pages from a large dictionary. This would shed light on the question of how representative of the whole English language the article is. Have the group omit abbreviations, symbols, place names, and repeated listings of the same word. One such survey showed 6-letter words leading with 24% (about 1/4 of all words) , followed by 8-letter words with 18% (2/9).

2. i-o. To make the fraction of 3-letter words, they must place the total number of words in the denominator (found by adding up words of all sizes) and the number of 3-letter words in the numerator. To make the decimal, divide the numerator *by* the denominator.

l. The chance of sticking a 6-letter word is the same as the fraction (decimal), found in **j,** of the whole article that is 6-letter words. The chance of *not* sticking a 6-letter word is just 1 minus that fraction (decimal). The decimals found in **m** or **n** should not be terribly far from the probability prediction **l,** but by chance they could be, for reasons given in **o**.

o. Probability is only a number tendency that groups of events approximate more closely as more experiments are performed. Any single event can deviate a great deal by chance. Furthermore, the second chosen article could, by chance truly have a different percentage of 6-letter words. The reasons students give should reflect these realities. By the way, it's much easier to compare the predicted probability and the experimental results if they are *decimals.*

CASE II: This is a fun little counter-intuitive experiment I've hit upon. In fact, the number of blues in the red pile and the number of reds in the blue pile will be the same each and every time! **There is a 100% probability of this. Throughout the experiments, you must not let on at all that you know anything about the anticipated outcome!**

Why this probability? Precisely because each pile still has the same number of counters *after* the color transfers have been done as *before.* But, in the final mostly-red pile, some of the reds have become blues. So where are the reds that used to be in the places occupied by these blues? Those are in the mostly-blue pile, replacing exactly the blues that are now in the red pile. The red "aliens" of the blue pile have to just equal the blue "aliens" of the red pile they replace! The whole secret to this simple solution is to de-focus from the exchange actions and simply study the state of the piles at the end. It takes several passes on this thought to really "feel" its truth.

Students may argue or fall off of this train of thought for a while before they Aha! the obvious truth of this, which is very pleasurable. But getting pumped up first in a theoretical argument is good for them! Putting their final theory in words is challenging and best done after a good group discussion. A group-approved written statement would be best.

 Mixing pop. This is a very challenging problem now made simple by the student's experience with counters. The average adult tends to complicate this problem a great deal. I've known some people to fill a page with algebra to solve it and others to fill the air with convoluted explanations! This problem leads to a good interpersonal experience with an adult and helps a student feel smart in the exchange!

CASE III: The predicted probability of throwing three dice and having an odd product is only 1/8! That's because an odd product occurs only when **all three dice** have <u>odd</u> numbers on them. There are only eight different combinations of evens and odds that can be on the first, second, and third dice: OOO, OOE, OEO, EOO, EEO, EOE, OEE, EEE. Anytime an even multiplies a number, the result is even. Thus the predicted amount of odd products in 80 throws is 10, while we expect 70 evens.

 Make it Fair. Most kids have a strong sense of fairness and will want to make this game fair. There are many ways they can creatively do so. Adding all three is the simplest way.

 Adding and Multiplying. The game is unfair because adding the first two dice, then multiplying by the third, will yield odd only 1/4 of the time (on the OEO and EOO combinations) of the eight listed above.

CASE IV. f. Other computable statistics might relate to the male-female count like the average number of males or females in cars, the average number of males and females in SUVs, or the fraction of SUVs driven by women. **h.** Other significant counts are the most popular color for SUVs driven by men, the most popular brand of trucks, etc.

i. The predictions about the next vehicle to come down the street will have the most chance of being correct if they are characteristics that are the most popular. For instance, since cars are most likely the most popular vehicle seen, and if the most popular brand of vehicle seen was Ford, the best bet might be that a Ford car would show up next.

Of course, overall statistics will never be a very good predictor of one individual datum. Statistics will simply create the most "wins" on your predictions in the long run. Therefore, expect most scores to be low. Students should discuss this concept as a class or in groups then formulate what principles are involved and how the scoring system might be revised.

Extension (Inter- and Intrapersonal): Ask the students to discuss in groups what they feel about **chance**. Have them choose one or more of the issues listed here and form a detailed, thoughtful group answer.

- Does *everything* happen by chance? That is, in a universe full of possibilities, don't some interesting things just *have* to happen?

- Or does *nothing* happen by chance? That is, maybe things that seem to just happen randomly really fit into a bigger order we have no knowledge of.

- Does it help for people to put a probability number on an event, like a 90% chance of rain, since the 10% means it still might not happen at all?

- Is there anything that has a probability of 1, or 100%, of happening (certainty)? Or is nothing above a chance of just .9999999999 or 99.99999999%?

- Do the things in dreams happen totally by chance, i.e., random connections between thoughts and images, or do they have a definite meaning, plan, or order?

- Even in random card games or flipping a coin do you feel like some people have *more* chance of getting good hands or a winning toss? Why is that?

Extension (Musical): Find a book or piece of sheet music that shows the notes of a familiar song.

- Do a tally of just the notes of the melody and see which note is used most often, which is used second most, and which is used least often in the song.

- Search for a connection between the most or second most-used notes and the words of the song that occur with these notes. Notice what words occur with the rarely-used note.

- Try to "compose" a tune on an instrument using only the first, second, and third most-used notes and occasionally the rare note. See if the tune has some of the "feel" of the original song.

- Find another song whose melody sounds quite different and repeat the tally. Are the most-used and least-used notes different?

AUTOMOBILE$

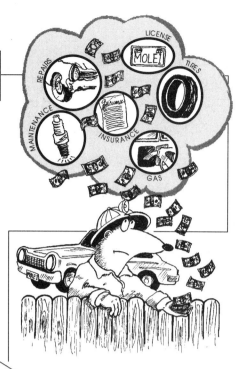

Cars are a big part of most people's lives in this country. There are over 160,000,000 of them in the U.S. alone! If, or when, you get involved with a car, you'll find out one thing for sure: they love to eat **money**! Many people don't bother to track how much money they feed their cars, so they waste a lot without knowing it. By carefully doing mathematics you can make better car and driving decisions in the future.

Choose your family car or a friend's car or truck as your math lesson. After you do your research, you'll be able to compute the **real** cost (not just for the gas) of driving any vehicle on a long trip, or just around town, even for a whole year.

The owner of the car should find your results interesting also.

 Here are some suggestions to keep in mind:

• Gathering this information may take you a few days.

• If you don't understand some of the words in the questions ask a knowledgeable person about them.

Be prepared to get some of the information from adults

• The vehicle's owner will already know some of the number answers, but probably not all.

• For some questions you may need to call a car dealer or mechanic. They will almost always be friendly and helpful, but if they aren't, try someone else.

• You may also need to consult the classified ads in a newspaper.

• Group planning and cooperation will help you also, because a partner may know ways to find things out if you don't.

• After you answer all the questions you will know car expenses from **a** to **z!**

Basic purchase information

a. Make of Car _____ Model: _____ Year_____

b. Type of car (e.g., van, sedan, pickup): _____

c. Year the car was **bought**. _____ **Cost** of car (without tax) when bought: _____ Sales **Tax**: _____

 helpful to do calculate write

d. Approximate **cost** of car when **new** (even if the owner bought it used): _____

e. Estimated **value** of the car now: _____

Fixed cost information

Fixed costs found in **f–h**, are what owners pay each year **even if they don't drive the car**. If the car is driven **only 1 mile** that year, the fixed costs are counted as part of the cost of driving that mile! All fixed costs are more for more expensive cars.

f. Find the approximate **decrease** in the value of the car each of six years after it was bought. You will use the blanks below.

The decrease in value for a new car should be greatest during its first year because people pay a lot for a new car and buyers want to lower the price a lot if a car is not new.

A car that's now **one year older**, with more miles on its odometer, has to be priced lower than the year before to tempt someone to buy it. This is called its **decrease in value** or its **depreciation**, which depends both on **age** and **miles driven**.

The car's decrease in value **this year** is part of the cost of driving the car this year. If a **lot of miles** are driven, the car has more wear and is worth less. Even if it isn't driven much, the extra year of **age** also makes it worth less.

Now **fill the blanks**. Ask the owner, check used car newspaper ads for prices of your car model at different ages, or call a used car dealer, then write the amount the car lost (will lose) in value each year.

1st year $_____	2nd year $_____
3rd year $_____	4th year $_____
5th year $_____	6th year $_____

this year _____

If the car is fairly new, use the blanks to predict its value decrease until its sixth year.

g. Cost of insurance for this year: $_____.

h. What is the cost of this year's license and registration for the car? _____

Driving cost information

i. What do four new **tires** of the car's type usually cost? _____

j. How many miles do the kind of tires on the car usually last?: _____ miles

Call a tire dealer or ask the owner.

helpful | to do | calculate | write

k. Known or estimated cost of **repairs** each year after purchase.

If the car is fairly <u>new</u>, repairs should be very low. As cars get <u>older</u>, or drive <u>lots</u> of <u>miles</u>, they need more repairs. Use the blanks to predict repair costs to the <u>sixth</u> year.

Repairs include all money paid to mechanics for parts and labor, including oil changes.

1st year $_____   2nd year $_____

3rd year $_____   4th year$_____

5th year $_____   6th year $_____

Estimate **this year's** repair costs _____

m. What does a quart of **oil** cost? _____

n. Quarts of oil added per **year:** _____ **cost**_____

o. About how much is an average gallon of **gasoline**? _____

p. About how many **miles** is the car driven per year?
_____ miles

q. Gas mileage. How many miles does the car go per gallon _____?

Find the **gas mileage** this way:

1) Fill the tank **full**.

2) Write down the mileage on the **odometer**_____

The mileage meter on the speedometer.

3) Wait until at least a third of the tank is emptied.

4) Fill the tank full again. Note how many gallons (and tenths of a gallon) were used: _____

5) Note the new odometer reading: _____

6) Using **2)**, **4)**, and **5)**, figure the **miles traveled** since the first fill and **divide** those **by** the **gallons** used.

Using your data to compute driving costs

Now you know enough facts to figure out what it costs to drive the car **a mile during this year**. (<u>Show</u> and <u>label</u> your calculations on another sheet of paper.)

r. Compute a year's tire cost _____

*Calculate this from **i**, **j**, and **p**.*

Hint: Estimate what **fraction** of the tires' life is used in **one** year. Find this fraction of the tires' **cost**.

helpful   to do   calculate   write

s. Compute a year's added-oil cost. _____ ← ... $\bigcirc$ *Calculate this from* **m** *and* **n**.

t. Compute a year's gasoline cost. _____ ← ... $\bigcirc$ *Calculate this from* o, p, *and* q.

u. Now add up all the costs of driving for the present year: **f**, **g**, **h**, **k**, **n**, **r**, **s**, and **t**._____

v. Use **p** with **u** to figure out how much it costs to drive one average mile this year. _____

Thoughtful questions about the cost of driving

← *Each of these takes some thought and time. Compute and label the work for these on a separate piece of paper.*

w. Pick a **city** you've gone to in a car or would like to see. Find out its distance and compute the *round-trip* driving cost to go there for a visit: _____

x. How many miles would your car have to drive before its driving costs would total as much as was paid for the car? _____

y. This question will show the effect of gas mileage on driving expenses. Compute carefully:

$\bigcirc$ *That is, if your car is a gas hog or a gas hummingbird, how will that change your driving costs in a year?*

- What would this year's driving costs be if the car got only **12** miles per gallon of gas? _____

- What would they be if the car got **40** miles per gallon?

  _____

- Is it worth it to buy a car with high gas mileage?

  Comment: _____

  _____

z. What was the cost for driving **a mile** during the **first year** the car was owned? _____

$\bigcirc$ Use the tax from **c** and the first-year info from **f** and **k**. Assume all the **other costs** were about the same that first year, unless you know some were very different.

The first year's cost includes the **tax** *paid on the purchase, because the tax will never be repaid even if the car is sold immediately.*

Your thoughts and comments. Write a paragraph, or discuss with your group, about how your attitude about cars has changed, if at all, since doing this Activity. Tell whether you would like to own a car and what kind you would drive, now that you know more about expenses.

helpful tips | to do | calculate | write

Teacher Notes

Intelligences Emphasized: Interpersonal, Intrapersonal, Bodily-Kinesthetic

Math Skills: Real-life data gathering, estimation, financial concepts, payments, rates, ratios, gas mileage, real-life story problems

Level of Challenge: Challenging to fairly difficult — involves using, obtaining, and understanding technical data

Materials: A vehicle whose owner is known by the student, calculator

Overview: This activity involves several new concepts, then hunting and gathering the myriad numbers that go with those concepts. The concepts all relate to automobiles and the seemingly simple question, "How much does it cost to drive a car a mile?"

This generates some complexity. After gathering data, the students must decide which operations are needed to compute useful information from them. The activity encourages mathematical interaction and discussion with adults about real information. The adults may arrive at information they never tracked before, thus showing students that careful math even has new things to show adults, who supposedly know all about their car economics. This Activity can also impart a sobering lesson in automobile economics to future drivers.

Introduction to Class: Discussing meanings of concepts like registration, insurance, gas mileage, and depreciation before the activity can make it go a lot smoother. On the other hand, explaining little and sending them off to obtain explanations and data from knowledgeable adults has a benefit also. It depends on your purposes.

If you choose the former, try some of these seeds for classroom discussion that help capture emerging student understandings:

• "How much do you think it costs to drive a car from here to (neighboring town)?" [Discussion]

• "What contributes to those costs?" [Lists responses on the board.]

• "Have you left anything out? For instance, what about tire wear?... Is that free?... And what about the fact that the car will then have more miles on its odometer, won't it be worth a little less when it's sold?... What about the wear and tear? Doesn't every trip mean that repairs and replacement parts will be needed sooner?... What are typical repetitive replacements and repairs?...

• "What about car insurance, a license, and *depreciation* (meaning that just the fact that the car is a year older it is considered less desirable and could sell for less)?... These are called *fixed costs*, meaning that you have to pay them whether you drive or not. Is *this* trip more expensive because of them, or should we say that this trip is part of all our driving for the year which is made more expensive by fixed costs? (The Activity includes fixed costs in the per-mile cost of driving.)

• "Finally, what about driving on the highway versus driving in the city? Which costs more and why? [Note: If you live in a city, car insurance is higher and stop-and-go driving is more wearing on a car. Gas mileage or efficiency goes down with city driving.]

• "You will find out what it costs to drive a particular car to (familiar town). You will need to gather a lot of information first from your parents or the car's owner. You might need to make a phone call to a used car dealer, or look at used car prices in newspaper ads to find out about your car depreciation. You may need to call a tire store about tire costs and how many miles tires last. You may need to keep track of gas at the pump so you can compute gas mileage. How much does the gas cost? You have ___ days to find out all these things, fill out the paper, and do your computing."

Extension: U.S. Cars. How much money is spent driving all the cars in the U.S. each year?

12

WATT'S HAPPENING? ELECTRICAL ENERGY!

Introducing Energy

Energy is the ability to perform work. It appears in many forms and can be changed to other forms depending on the work to be done. For example, the sun sends light and heat energy into space. A small fraction of it strikes the earth and does the work of melting glaciers. The sun's light and heat can also be converted by solar cells and solar panels into electricity for a light bulb and hot water for a shower.

Water has moving or "kinetic" energy when it runs in rivers all over the earth. It's moving energy can be converted to electricity with dams. The electricity is then converted by motors to motion in a CD player and by special wires into heat in a hairdryer.

This Activity is about **electrical energy**. We measure it in **watts.**

 a. Hold your tennis shoe (about 3/4 pound) in your hand. Lift it **one foot** higher in one second (not instantly, but for the full time of **one second**). Practice until you can take a second to lift it one foot. Each time you are using about a **watt** of energy.

The horsepower

It takes **746** of these watts of energy to give a motor **one horsepower** of energy. A horsepower is the amount of energy it takes to lift *five* middle-schoolers (550 lb.) *one foot* off the ground in *one second.* Many things in or near where you live require far **more** than a horsepower worth of watts to run. For instance the heat and spinning energy it takes to begin running a **clothes dryer** can be 5222 watts.

 b. How many middle-schoolers (at 110 pounds each) could be lifted a **foot** in **one second** by the dryer's energy?___

 c. On a blank sheet of paper draw a picture of a hot clothes dryer lifting a large platform with a rope. Place the right number of middle-schoolers on it.

 d. How many **feet** could **five** middle-schoolers be lifted in **one second** by the clothes-dryer's energy working in a motor?

*If something is going to move at all, it needs some form of **energy**. A car gets it from gasoline, propane, or electricity, a plant gets it from the sun, and you get it from food.*

Watts are named to honor James Watt, the Scotsman who designed some of the first steam engines in the late 1700s.

Picture this happening in your mind as you say fairly quickly "one thousand one."

*Think about the answer to **b**; it's related to this one.*

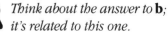

 helpful   to do   calculate   write

Power

If a certain amount of energy, like 500 watts, is used for a period of time (like an hour) to do work, scientists say that 500 "watt-hours" of "**electrical power**" is being used. Each watt being used for an hour is a **"watt-hour"** of power used.

 The electric company charges us for every watt-hour of power they send through the lines to our home. A horsepower of watts (746) being used for two hours would mean an electric company charge for 2 x 746 = 1,492 watt-hours.

When we have something powerful, like a hot-water heater, that requires 5,000 watts of energy to make it work, power companies prefer a unit called the "**kilowatt**," which means **1,000 watts**.

It takes about **5 kilowatts** to start up a hot water heater (i.e., it is a 5-kilowatt appliance). If it is used for seven hours, it consumes 5 kilowatts x 7 hours = 35 kilowatt-hours of power.

 a. How many kilowatt-hours of power does it take to run a 1500 watt space heater for 12 hours? _____
Two 100-watt light bulbs for 20 hours?_____

For *each* kilowatt-hour a power company can charge anywhere from 5¢ to 25¢ depending on where you live, how hard it is to find or generate energy, and how hard it is to get it to where you live.

 b. What are the **highest** and **lowest** charges from power companies for the use of the space heater _____, _____ and light bulbs _____, _____ in **a**?

 c. After thinking and discussing with others, fill out this list:

Three things that could make electricity low-cost in an area	Three things that could make electricity expensive in an area
1.	1.
2.	2.
3.	3.

helpful   to do   calculate   write

Energy At Home

Now you have a chance to experiment and learn a lot about the electrical energy that comes into your home and how it is used. To gather information to answer these energy and power questions you may need to check bills, make phone calls, or talk to adults.

 a. How much does a **kilowatt-hour (kW-hr)** of electricity cost? Find this on a bill if you can. _____¢ per kW-hr

 b. How many **kilowatt-hours** of electrical power does your home or apartment use in one month? _____

 c. Find your power meter, if your home or apartment has its own. Apartments may not, so for this you may need to look at the meter on the house or apartment of someone you know. A typical meter is shown at the right. Look at it carefully, even very close up. Something is moving, maybe very slowly. What is it? _____

_____

 A turning wheel means that electricity is flowing through the meter into the home **right now**. One *dial* may be slowly turning, too, but too slow to tell for sure.

 d. Each of the dials, from left to right, tells a digit of a number. Read the whole number carefully. **Read any print** near the dial — it may be an instruction to *multiply by 10* to get the number of kW-hours it is really reporting. **How many kilowatt-hours is the meter reporting right now?** _____

 Every month your power company sends someone around to read everyone's meters. They check to see how much the reading has changed from last month. The change was caused by how much electrical power was used this month.

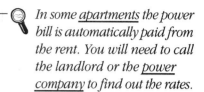
In some <u>apartments</u> the power bill is automatically paid from the rent. You will need to call the landlord or the <u>power company</u> to find out the rates.

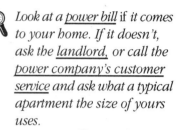
Look at a <u>power bill</u> if it comes to your home. If it doesn't, ask the <u>landlord,</u> or call the <u>power company's customer service</u> and ask what a typical apartment the size of yours uses.

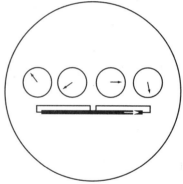

This number has been growing from the time the meter was first installed!

helpful | to do | calculate | write

A Meter Experiment

 DO THIS ONLY WITH AN ADULT TO SUPERVISE. It's best when almost nobody is at home and it's daytime, probably on a weekend.

a. Try to turn off all of the electrical things in the house. ◄—— *One way is to check every outlet plug and switch in the home.*

Ask people to cooperate for just a few minutes. Remember heaters, lights, clocks, kitchen appliances, stereos, fish tanks, TVs, dryer, answering machine, etc. Remember that a TV can be drawing electricity even when it's turned off, so unplug it. If there is a control dial for all the *electric* heat, turn that way down or off.

> **Refrigerator:** Leave the loose cord showing so you will remember to plug it back in. If you can't reach the plug, turn the dial inside the refrigerator to *off*, and put a **big note** in the kitchen reminding you to turn it back on. <u>Don't leave it off or unplugged — it will thaw the freezer!!!</u>)

◄—— *Also remember the <u>refrigerator.</u> Best to unplug it, if possible, <u>five minutes at most</u>.*

 b. Has the meter stopped? If it hasn't, and the wheel is turning pretty fast, most likely the **hot water heater** is on.

> **Hot Water Heater:** It's the hardest thing to turn off. There's no switch or plug. Try to turn it off with **adult supervision** at your central electrical switch panel, where there should be a marked switch for the water heater. After the switch is thrown it may need to be thrown *both ways* to make it turn on. (*Don't leave it off for long either!*)

 Now the meter should be stopped. If it's moving very slowly, there may be something you forgot.

 c. Try to find anything you forgot and turn it off. If you can't find it, turn off each switch in your **electrical panel** (with adult supervision) one at a time, until the meter stops completely.

helpful   to do   calculate   write

d. The experiment. Write a number in each blank that tells how fast or slow the wheel in the meter is turning when **only** that item is turned on! (1 = very slow, 5 = very fast, with 2, 3, 4 in between).

small light _____   dryer _____
large light _____   electric stove burner _____
TV_____   microwave _____
stereo_____   water heater _____
hair dryer_____   space heater _____
wall heater _____   stereo_____
(other)_____   (other)_____

Do this only for the things you have. Use "other" to try some things not listed.

e. Go back, plug everything in again, and turn on all the panel switches.

f. What made the meter move slowest? _____ Second slowest? _____ What made it move fastest? _____ Second fastest? _____

g. Make up another question that you can answer from the experiment. _____
_____.

And your answer: _____

The Household Electrical Chart

a. Fill in the first two columns of the **Household Electrical Chart.** In them list 10 or more items that use electricity in your home. Then do a "watt-search" to find out and record how many watts each of these uses.

You need to become a real detective to find the watts for them all!

Be sure to include the **water heater**, **stove**, **refrigerator**, and **electric heaters** if they use electricity. Below the **Chart** are many valuable **hints** about how to "watt-search" on different electrical items.

These use tons of electricity!

b. In the third column, estimate the number of hours **last month,** that you think the item was used.

If last month was not typical, choose another month.

c. Multiply **watts** by **hours per month** to get the watt-hours per month for the next column.

helpful | to do | calculate | write

d. Divide each answer in **c** by 1000 to get the **kilowatt-hours per month** the item uses (round to two decimal places) for the fifth column.

The quickest way to divide by 1000 is to move the decimal point at the end of your number back three places.

e. For the last column, compute **how much it costs** to operate each item per month by using the kilowatt-hour cost for your area.

f. Use the Chart to estimate the **total monthly cost of electricity for your home for one month.** _____

g. Compare your estimate with the electric bill, and comment on how they differ. _____
_____.

If they're very different, try to figure out why. Discuss what you found with an adult who may be able to see what you did wrong or left out.

h. Use the Chart to make up some questions that you could answer from it. Examples: "What is the cost per year to use a refrigerator?" or "What is the cost for all lights in a month?"

_____
_____
_____

helpful   to do   calculate   write

Household Electrical Chart

ITEM	NUMBER OF WATTS IT NEEDS TO START UP	NUMBER OF HOURS PER MONTH IT IS ON	WATT-HOURS PER MONTH IT USES	KILOWATT-HOURS PER MONTH IT USES	COST PER MONTH FOR RUNNING THIS ITEM

 ### Helpful Hints for Your Watt-Search

 Lights are easy — just read the top or side of the bulb(s).

 Thoroughly search each **appliance**, usually its back or bottom, until you find how many **watts** the manufacturer says it uses. This information should appear on a sticker or tag or in a rectangle molded in the plastic. You just have to find it. It might have "110 volts", "2.6 amps" (if you multiply volts and amps it will give you the watts), but somewhere it will usually tell how many **watts** it uses. (Just a "W" means watts.) If it just gives "1.6 amps", multiply by 110 volts to get watts.

 Water heater watts are usually shown somewhere on a sticker or metal plate.

 Refrigerator watts may be inside the door, on the door, on the back, or behind a removable panel under the door. Otherwise call an appliance store and ask.

 Electric stoves can be tricky. Try looking for a metal plate just inside the oven door. Otherwise it may be on the back and impossible to see. If you can't find it, call an appliance store and ask the watts for a stove like yours. Stoves have many different amounts of energy they use depending on whether a low burner is on, several burners are hot, or the oven is on. Try to find out the watts used for these different cases.

 Electric heaters may be tricky if they're mounted in or along the wall. Look for a plate or stamp in the metal. Otherwise call a hardware or heater store, or ask a knowledgeable adult, to find out typical watts for these.

Teacher Notes

Intelligences Emphasized: Kinesthetic, Interpersonal, Linguistic, Spatial

Math Skills, Concepts: Electrical units, simple computation with units, tabulating and computing with real life data.

Materials: An actual residential electrical usage meter, calculator

Level of Challenge: Upper intermediate to middle-school, moderately challenging

Overview: This activity immerses students in the workings of math in real life via electrical power. They learn what it is, what uses it, and how rapidly, and how it is paid for. There is quite a bit of information on the technicalities of measuring power. It requires that they have access to the exterior electric meter at home or at a friend's or relative's place. The meter must be one that measures only the electricity used by one residence.

The experimentors are then guided through a series of activities with the meter and with real objects in the home that lead them to a better appreciation of electrical power. It is my opinion that such knowledge is essential for youths to understand energy issues and thereby become more responsible users and stewards of energy as adults.

Introduction to Class: Any number of lead-ins are possible, but I'll propose this one. Ask the students where household electricity comes from. After answers emerge that include generators, wires, hookups, and outlets, you could ask whether or not it's free. What makes it costly to generate and distribute power? If it's not free, then how do we pay for it? How does the power company, located far away, know exactly how much you are using when you dry your hair? You have many neighbors whose electricity comes through the same wires across the landscape. (A field trip to a power plant before, during, or after this unit would be very valuable.)

Then pass out only the sheets with the **Introducing Energy**, and **Power** sections on them. Ask students to discuss the ideas of electrical energy and power in groups and fill in the sheets. The next two sheets must be done as an outside assignment, preferably including a weekend. The **Energy At Home** section can be done individually, then, depending on each student's access to an electric meter, **A Meter Experiment** can be done by a small group at one

The "PM" on the mole's costume stands for "Power Mole"!

student's house, followed by a group report on the results.

The last two sheets, **An Electrical Chart**, are best done individually at home, possibly requiring a couple of days. Parental input on it would be helpful.

Answers

Introducing Energy. b. 5222 watts that activate a dryer, divided by 746 watts per horsepower, makes *seven* horsepower when used in a motor. Multiply this by five middle-schoolers to get 35 kids being lifted *a foot* by the dryer's energy in *one second.*

d. Seven. This is just a translation of **b**. Some students may have to think about this one a while. The seven horsepower can lift 35 middle-schoolers one foot or *five* middle-schoolers *seven* feet in one second. Both do 35 "middle-schooler-feet" of work.

Power. a. 18 kw-hr, 4 kw-hr **b.** space-heater: $4.50 to $.90; light bulbs: $1.00 to $.20 **c.** The reason why power can be low-cost is it can come from lower-cost sources, like dams in the Northwest. (This low-cost power is proving environmentally expensive with the loss of salmon in the rivers that close down fishing and tourist industries.) Electricity in cities can be cheaper than in rural areas where each customer needs more wire over sometimes difficult terrain to be reached. Solar and wind energy, if very available in an area, can be medium-high in cost until these newer facilities have paid themselves off. Electricity can be costly if it is generated in fossil-fueled generators or if the demand for it is very high in an area due to factories and many homes.

The Household Electrical Chart. The students may find that their electric bill looks a lot different than their calculated monthly use of electricity. This should mean that they have overlooked major costs like baking in an electric oven, computer use, electric blankets, hot water use, etc. Ask them to discuss any major discrepancies and the possible reasons for them. There's always the chance that the power company may be overcharging them!

Extensions:
(Linguistic, Spatial) Electricity Report. Write, tape, or draw a report about electricity — what it is, how it is generated, and how it gets to homes.

(Intrapersonal) What do you really need? List the electrical items in your house in decreasing order according to how necessary to you they feel. To do this, imagine that only *one* electrical item was allowed in each home. What would you choose? How about *two* items? etc.

FROM NUMBER NUMBNESS TO "FEELING" FIGURES

Numbers are found almost everywhere: in TV and radio news, programs, commercials, school books, magazines and newspapers. We hear and see so many of them, and they seem so big or so small, we can't "feel" their meaning. Most people just let them float by. What do *you* do?

 a. To find out, here are some important facts with numbers in them. Go ahead and read them.

- On average the **sun** is 93,000,000 miles from the earth.
- There were 241,000,000 **cars, trucks and buses** in the U.S. in 2005.
- Each year 400,000,000 gallons of **motor oil** from American cars are wastefully dumped, rather than recycled, causing pollution of ground and water supplies.
- In 2006 in the U.S., 28 billion pounds of **plastic packaging** were thrown away.
- The F-22 is a very high technology U.S. **fighter aircraft**. One F-22 fighter costs $360,000,000.
- A **germ** (bacteria) is about .0002 inch long. A typical virus is .0000004 inch long.

 b. Did most of the numbers "fly by" or did most "click" in your mind? _____

Feeling the figures

 c. Without looking back, which number fact do you remember best? Write it here:

_____

 d. Looking back, which number fact was the most striking to you? _____

 e. Looking back, which **two** number facts didn't mean much to you?

_____

_____

 helpful to do calculate write

Here are some of the **same** facts, but they've been made more **meaningful**, so you can "feel" their size. Read them.

- If you drove to the sun at 60 miles per hour it would take **177 years** of solid driving — 24 hours a day — to get there! You would've needed to start driving about the year 1820 to arrive there today!
- If all the cars, trucks and buses in the U.S. were placed **end to end**, they would circle the earth over **29** times! Or they would make a bumper-to-bumper traffic jam all the way to the **moon** and back, and then there again!
- The plastic packaging thrown away each year in the U.S. **weighs** as much as **77,400** Boeing 747-400 Jumbo Jet airliners. These airplanes would line up nose to tail for 3400 miles, or across the U.S. from Seattle to Miami!
- The cost of an F-22 fighter would make a string of dollar bills, attached end to end, that circles the earth once and 2/5 more.

 f. Are these numbers more surprising or amazing now that they are easier to see or feel? _____

 g. Which fact struck you the most? _____

 h. What feeling or thought or image came to you as you read it? _____

Now It's Your Turn

You saw the number facts get more **powerful** thanks to some mathematical figuring. Almost any real-life numbers can be made more striking with math. This Activity will show you how by having you practice on some facts mentioned earlier.

> Each year 400,000,000 gallons of motor oil from American cars are wastefully dumped, rather than recycled, causing pollution of ground and water supplies.

First think: What would make a lot of liquid like that meaningful? What could it compare to? Try answering the following questions to get an idea.

 a. Name four situations, both human-built and natural, where you have heard of a lot of liquid in one place (for example, a pond). _____, _____, _____, _____.

 helpful to do calculate write

b. For which of these, if any, could a student find a **measurement** of how much liquid they can hold?

_____

For now, unless you look up one of your own suggestions, these figures can help you:

- An average-sized supertanker holds **100,000,000** gallons of oil.
- A typical large swimming pool holds about **600,000** gallons of water.

 c. Compute and make **two** meaningful statements about how much motor oil is dumped wastefully each year:

_____

_____

> A germ is about .0002 inch long.

 d. Measure five lengths, as a fractions of an inch, of *short* everyday objects. Example: "pinkie fingernail ³/₈ in." :

_____ ___ in., _____ ___ in.,
_____ ___ in., _____ ___ in.,
_____ ___ in.

 e. Choose one of the objects and its length as a **fraction** of an inch. Change that to a **decimal part** of an inch. **Divide** the object's length **by** the .0002 inches of germ length.

 f. Write your meaningful statement about germs here:

_____

_____

> In 2006, 11,920 murders were committed with guns in the U.S. In Japan, with 2/5 of the U.S. population, there were only 35. In Britain and Wales, with 1/5 of the U.S. population, there were only 23.

This unsettling number (and the small numbers compared with it) are fairly striking already, but they're hard to imagine. One thing is certain — every murder produces a dead body that has a certain size. This fact can lead you to some meaning for the yearly number of murders.

This length is hard to imagine. To "feel" it we could compare it to a small length that everyone knows.

*CALCULATOR TIP: A **fraction** of an inch, like ³/₈ inches, can be put into the calculator as a **decimal** part of an inch by dividing the **numerator** of the fraction, 3, by the **denominator**, 8, on your calculator: 3 ÷ 8 = .375 in.*

Questions to consider:
- *How much **area** does that many bodies (or their caskets) take up?*
- *If the caskets were lined up, how long of a line would they make?*
- *How much room would the people have taken up when they were still alive and standing?*

helpful   to do   calculate   write

If you choose the **area** idea, you need a standard well-known area, like a **basketball court** or **football field**, to compare with the room needed for all the bodies and caskets. Here are some researched facts and some approximations:

- Official high-school basketball court: 84 ft. long by 50 ft. wide.
- Football field: 360 ft. long by 160 ft. wide.
- General casket size: 7 ft. long by 2 ft. wide.
- A standing person takes up a square about 2 ft. long by 2 ft. wide.
- A lying down body (on average) takes up about 7 square feet.

 g. Use these facts to create **three** meaningful statements or images to help better imagine the number 11,920 of U.S. murders compared to the small numbers of murders in Japan and England and Wales.

Write them on the back of this sheet or another paper. Label your work and show the steps you took.

Make more numbers meaningful

Here are some real number facts (and some suggestions for making them meaningful). After doing your research and figuring, turn each number fact into a startling statement.

> In 1996, the heaviest living person was a man who weighed 891 pounds, and measured 120 inches around the chest, 116 inches around the waist, and 70 inches around the thighs.

 Find objects that are about 70, 116, and 120 inches around, and another object that weighs about 900 pounds, then compare the man's measurements to them in statement **h**:

Objects that are cylinder-shaped are better for comparing the measurements. Any shape of object is fine for the weight.

 h. "The heaviest living person in 1996 weighed as much as _____ , had thighs as big around as _____, a chest as big around as _____, and a waist as big as _____, and thighs as big as _____.

> The 2007 population of the United States: about 302,000,000.

 Find, using a ruler and a scale on a map, the **longest distance** across the United States (Washington to Florida) or the distance from the western tip of Alaska to the southern tip of Florida.

helpful | to do | calculate | write

Imagine that all of the people in the U.S. stood tightly packed in a line, taking up **one foot** of length per person. Compare the line of people with the distances you found on the map.

 i. Make your statement about the U.S. population:_____

_____

_____

> The greatest weight ever lifted by a man was 6,270 pounds in 1957. He pushed up with his back in a special device to do it.

 Compare the weight he lifted to something you can find the weight of: an elephant, a car, a schoolbus.

 j. Your statement:_____

_____

_____

You can also use <u>many</u> smaller weights, like adults or kids.

> The fastest car (rocket-engined) was driven in 1979 at a speed of 740 miles per hour.

 How long, at this speed, would it take to go a short or long distance you know well?

 k. Your statement: _____

_____

You can compare the speed better to something you know if you convert the <u>miles per hour</u> to <u>feet per hour</u>, then to <u>feet per minute</u>, then to <u>feet per second</u>.

> Most scientists believe the **age of the earth** is 4.6 billion (4,600 million) years old. Other events most scientists have estimated are:
> **Start of life:** 570 million years ago
>   (4,030 million years from the beginning);
> **First mammals:** 270 million years ago;
> **First bird:** 195 million;
> **First Tyrannosaurus Rex :** 141 million;
> **First woolly mammoth:** 1.8 million;
> **First people:** 1.5 million.

Each date means <u>time from now</u> ("years ago"). Round them off to the nearest 100,000,000 years, then find their place back from the present.

 l. Draw a time line 138 inches (11 1/2 feet) long for the estimated 4.6 billion year life of the earth up to now. This allows you to mark off **3 inches** for each **100 million (.1 billion) years** of the estimated age of the earth.

helpful   to do   calculate   write

Pretend that your line is like **one year**, and carefully mark, with another color, **twelve** 11 1/2 in. spaces representing the 12 "months" of the earth's "one year" life. Name each "month" with the regular month name.

 m. Figure what "dates" in the earth's "year" each of the important events occurred (e.g. December 10 for the First bird) and mark them on the time line. Label them.

> On the average, every mouthful of food eaten in the United States has traveled 1300 miles.

 If every mouthful travels 1300 miles, so does your whole meal! **Everyone's** meal moves 1300 miles to their plate! A possible way to "feel" this number is just to look up places on the map that are 1300 miles from you and refer to one of those places in your statement about food in **o.** 1) below.

 Another way get the impact of this fact is to use the figure of 3 pounds of food eaten per day by the average person in the U.S.

 n. Compute the large number of tons of food moving 1300 miles every day for all 273,000,000 Americans, and relate it to the 22 tons that can be hauled by a semi-trailer truck.

 o. Write two statements that make the food fact meaningful:
1) Every mouthful of food eaten in the U.S. has travelled the equivalent of from _____ to _____.
2) The food eaten every day by Americans requires transportation equivalent to _____ large trucks driving it 1300 miles.

 On Your Own. Now that you are more experienced, try making one or more other number facts meaningful. Show your **figures and calculations**, with **comments** on your steps, in order to back up your statements!

Once you choose a number fact, and get a neat idea of how to make it meaningful, you will need to research **more** numbers to compute your idea. Good references for number facts are books like *The Guinness Book of World Records*, *World Almanac*, and *Information Please Almanac* (www.informationplease.com).

CALCULATOR TIP: *Use the figure 273,000,000 for the U.S. population and compute how many tons (1 ton = 2000 lb.) of food have to move 1300 miles for all Americans' meals in one day.*

CALCULATOR TIP: *Leave off 3 or 6 zeros to put a big number into the calculator and prevent overflow. Then place them back on the answer (or move the decimal point three places to the right) to make your answer the right size.*

Search for an interesting number in newspapers, magazines, textbooks, or the internet.

 helpful to do calculate write

Teacher Notes

Intelligences Emphasized: Intrapersonal, Spatial, Linguistic, Interpersonal

Math Skills, Concepts: Large numbers, fractions, decimals, calculator skills, estimation, using and converting a variety of measurement units, areas, volumes, charts, tables, researching real-life data

Materials: Calculator, research sources, ruler

Level of Challenge: Moderately challenging to very challenging

Overview: This is perhaps one of the most important of the activities for each student's future as a citizen. The numerical information we are all bombarded with daily is often embedded in text or charts and has strange sounding units, like "pounds per cubic foot." Very large or very small numbers are used, which are hard to comprehend, even for adults. A million or billion of something is a concept but not a visual experience. With youngsters, whose life experience is less than ours, data is even more difficult to grasp. They simply opt to let statistics fly by them with little meaning attached.

Journalists know about "**number numbness,**" so they often try to make a statistic meaningful, i.e, "The amount of water needed to raise one steer can float a battleship." Developing this image requires knowing how much a battleship weighs and how much water is used for feeding, growing the grain for, and washing one steer. It also requires knowing how much a gallon of water weighs. All this also requires **research** followed by calculation with estimated and rounded figures.

This Activity introduces youngsters to the fine art of making numbers meaningful, a skill that can be refined and used for a lifetime of communicating number ideas. The students need to be psychologically prepared for a mathematical workout requiring calculator, good visualization skills, imagination, research skills, and verbal abilities. They'll be given some interesting statistics, with some of them already "made meaningful," and others supported with detailed information on how to bring them to life. As the Activity progresses, students are required to think out and find more of the details for their work.

The students are then encouraged to do the same with statistics found on their own. This will involve some research (sources are suggested in the Activity) to find the statistic, then some imaginative mental play to relate it to something they know, then *more* research to find the numbers that will convert the statistic to a meaningful idea. *One* statistic can become quite a project!

This Activity should not be thought of as something to be "finished." Any part of it can suffice as a lesson or class activity. The bulk of it would ideally be done in several exposures over several weeks involving time for lessons on the units and principles involved as well as for doing the project. The first explorations students do on their own may require encouragement and guidance in visualization and research.

Introduction to the class: Ask the class if they know any interesting *real life* numbers. Most may be blank at first until you suggest a few, or some students contribute them. Point out that numbers swirl about us all the time. Have the class brainstorm a list of sources that pitch them at us.

Then group the students, preferably, and hand out only the first two pages of the Activity. The first part is intrapersonally oriented, soliciting thoughts, feelings, and reactions. Encourage discussion on those before students answer.

Go no further than the third page in the first exposure, announcing that the students are about to encounter a lot of interesting real-life numbers.

Remember: Only a few questions from this Activity should be attempted in any one time block to avoid superficiality of understanding or fatigue. Since the pages and problems generally grow more challenging as they go you can encourage more able students to go further and deeper into ideas and calculations.

A good stand-alone follow-up assignment is to ask the students to go for a statistic search. You can even have a contest for the five most unusual, worthwhile, surprising, or mathematically interesting statistics!

Answers

Now It's Your Turn (Note: These answers are only one way of computing and expressing the results based on information discussed.) **c.** Something like "Four supertankers of motor oil are dumped wastefully each year" or, "The amount of oil wastefully disposed of each year would fill 667 large swimming pools; that is, almost two pools per day are filled with waste oil."

f. Something like: "1,250 germs can be lined up across a worn-down pencil eraser."

g. Students may offer different or less complete versions of this full statement of the situation: "The number of caskets filled with people shot by guns in 2006 in the U.S. would cover almost 3 football fields, or 40 basketball courts. Those shot in Japan (2/5 the size of the U.S.) would fill only 1/13 of a basketball court instead of a proportional 17 courts). Those shot in England and Wales, 1/5 the size of the U. S., would fill 1/9 of a basketball court (instead of a proportional 8 courts)."
(Discuss …)

The answers to **h - o** will vary widely depending on student choices. Students can check each other's conclusions to make sure they are calculating right.

Note: Following each statistic-made-meaningful are many openings for mini-lessons on science, environment, and social studies that can be productive of thought and awareness. A major opening comes with the attempt to interpret the statistics by seeking their causes.

For instance, the gun statistics are fairly shocking. It can easily arise that students want to know why the numbers are proportionately so high for the U.S. This gives rise to examining many factors ranging from the rise of inner city gangs, to the powerful political role of the National Rifle Association and its resistance to gun control, to how guns are portrayed as solutions to problems in our media. It also gives an opening to seeking factors and gun control measures in British and Japanese society that work against extensive gun use.

14 GET REALLY SMART
CRACK PROBLEMS WITH ALL YOUR INTELLIGENCES

Did you ever think you might actually have more than one brain, or at least, more than one way of being smart? Do you know that intelligence is not just the two things many people think of as smart: being good with words and sharp with logic? In fact, there are at least **eight** intelligences that every person can use and some of them are your main ways of being bright. Then why not solve brain-stretchers using the cleverest parts of your mind that often include more than words or logic (though they're OK too). The brain-stretchers you will be working with will draw on your **other intelligences**, so be sure to put them to work.

 a. Read the list of things you can do while solving tricky brain-stretchers. It's on page 14-2 called **Multiple Intelligence Strategies**.

 b. Play the **Timed MI Sort Game** with partner(s). It's based on the **Multiple Intelligence Strategies** sheet. To play it,

1) draw **eight** 3 inch-by-4 inch "boxes" on sheets of paper and 2) make **24** cards with one of these words or phrases on each card:

Linguistic	Imagine scene	Act Out
Spatial	Track feelings	Make Chart
Bodily-Kinesthetic	Moving objects	Join Groups
Interpersonal	Sense pattern rhythm	Diagram
Musical-rhythmic	Exact meanings	Own Words
Intrapersonal	Report results	Sort
Mathematical-Logical	Watch thinking	Systematic
Naturalist	Estimating before calculating	Classify

RULES

A **turn** for a player consists of:
- being given a **shuffled** "deck" of the cards by another player, and putting the deck blank side up on the table;
- drawing one card at a time from the deck and placing it in one of the **eight boxes** on the paper. After <u>all</u> cards are drawn, **all the strategies** must be placed correctly with each of the **eight intelligences** in the eight boxes.
- That player is **timed** by another player as he or she does the sorting task. Each player takes **three turns**. The times for all turns are added up. The player with the **least total time** wins.

 Example: If a player draws "Exact meaning" and has not yet drawn "Mathematical-Logical," that card is placed in a blank box to be joined by "Mathematical-Logical," later.

 helpful to do calculate write

Multiple Intelligence Strategies

 Let your *Linguistic* or language intelligence help you put what is being asked into **your own words**. That intelligence also helps you **report your results** afterwards so you and others can understand what you did.

 Your *Spatial* intelligence may click on a method of solving a brain-stretcher if you make things visual by drawing a **diagram**, **imagining** a detailed **scene** in your mind, or making a **chart** that shows how parts of the brain-stretcher relate.

 You can turn your *Bodily-kinesthetic* intelligence into a helper by **moving objects** (or your hands) around to represent the brain-stretcher, or you can **act out** something from the brain-stretcher with your body.

 Use your *Interpersonal* intelligence to help you **join groups** and consult with **others** to make progress — the strongest intelligence of each person in your group can contribute to the solution.

 If your *Musical-rhythmic* intelligence is tuned in, you can sense a **rhythm** or **pattern** in the brain-stretcher that leads to the solution.

 Use your *Intrapersonal* intelligence to help you closely **watch** how your **mind thinks** and to catch negative thoughts and **feelings** that may cause you to get stuck.

 Use your *Logical-mathematical* intelligence by:
- being very **organized** and **systematic** when you're checking an idea,
- paying careful attention to the **logical** meaning of each word and phrase in the question,
- using your knowledge of the **meanings** of adding, subtracting, multiplying, dividing, and other mathematical methods, to find the right tool, and,
- **estimating** what the answer might be before you get into calculating.

 Use your *Naturalist* intelligence by looking for subtle details, likenesses and differences in your data and by sorting parts of the problem into different groups or cases that you name.

Working the Brain-stretchers

In the list below is some basic advice to help you get going on the brain-stretchers, one at a time. The suggestions will help to keep you in a positive frame of mind while you puzzle each one out.

 c. Though you are eager to solve your first brain-stretcher, carefully read the advice below first. The best way of all is to take turns reading each suggestion aloud in a group. When you are **stuck on a problem**, re-read the first five suggestions. When you think you are **finished**, re-read the last three suggestions.

ADVICE FOR WORKING THE PROBLEMS

- **Don't expect to get the answers fast**. Plan to work on each one for **quite a while** before you get the solution and have it in a presentable form. This will require **patience** and **perseverance** as a math "detective."

- **Shop first for a less difficult brain-stretcher** near the beginning of the brain-stretcher set that appeals to you. This is a good way to get **warmed up** before tackling the later, harder ones.

- **Spread the brain-stretchers out.** That is, take a few days to do them, working on only **one or two each day**. You'll pay closer attention to each one.

- **If you're stuck on a brain-stretcher,** try bringing in **another intelligence** by looking at your list of **Multiple Intelligence Strategies**. If necessary ask your teacher more about how to use your strongest intelligences on that brain-stretcher.

- **It's OK to get temporarily bogged down** on a brain-stretcher. Sometimes a solution will occur to you after you **let it rest** then **return** to it. Your teacher may have some **hints** to give you if you're still stuck after **24 hours**.

- **Use a separate sheet of paper for each brain-stretcher.** Show your **thought work** on it: your calculations, diagrams, scribbles, dead ends, and successes. Try to keep your tries **organized** and **labeled** rather than as random scratchings on several pages.

- **When you find a solution, write some sentences** describing the solution and how you found it. You can use **diagrams or charts** to help make your explanations clearer. If speaking or drawing is easier for you, arrange to make an **oral report** to your teacher or the class.

- **If you didn't get a solution,** write or report about exactly **how you *didn't* get it**: show what things you tried and how they **didn't** work out.

Brain-Stretchers

1. A girl has a lot of rabbits she keeps in hutches. She plans to sell them when they are grown. Someone asks her how many rabbits she has. She says "I don't know, but when I put **nine** in each of my hutches, **one** is left over. When I try putting **eleven** in each hutch, **one whole hutch** is left empty." How many **rabbits** and **hutches** does she have?

 If another girl puts 15 of her rabbits in each hutch there is **one** rabbit left over. If she puts 17 in each, she has **one hutch** left empty. How many rabbits and hutches does she have? Answer this then try to make up and solve a similar rabbit brain-stretcher with two other numbers in it. (Careful! Not just any two numbers work out.)

TRY OTHER INTELLIGENCES • • • SHOW YOUR THOUGHT WORK • • • REPORT YOUR PROGRESS

2. A father takes a walk with his daughter. The daughter takes **three** steps for every **two** the father takes. They both start across the street on their **left** feet. How many steps will each take **after that** until both step together on their **left feet** again?

 Work the same brain-stretcher for a daughter who takes **four** steps as her father takes **three.**

TRY OTHER INTELLIGENCES • • • SHOW YOUR THOUGHT WORK • • • REPORT YOUR PROGRESS

3. Five people, all of **different weights,** have an **average** weight of 173 pounds. What might each person's weight be? Then give three other possible answers (sets of five weights each) for this question.

TRY OTHER INTELLIGENCES • • • SHOW YOUR THOUGHT WORK • • • REPORT YOUR PROGRESS

4. Here's a simple game called "**Even-odd**" for two players. First find two regular **dice** and a piece of paper for a scoresheet.

RULES
- One partner decides to be "**Odd**" and the other partner decides to be "**Even**."
- Each partner throws one die. **Multiply** the two thrown numbers.
- If the answer (the product) is odd, "*Odd*" wins. If the answer is even, "*Even*" wins.
- The player with the **most points** after 20 dice throws wins.

a. Before you play, does this game sound fair to you? yes no

b. Play the game.

c. Does this game seem **fair** to you after playing? yes no

d. Why? _____

e. You might have just given an **opinion**. Now **analyze** the game on another sheet of paper until you can give real **evidence** for how many times you would expect *Even* to win and *Odd* to win in 100 throws.

Explore how fair these games are. First **predict** whether they're fair, **play** them, then **analyze** them:
1) **Even** wins when the sum of the dice is even; **Odd** wins when the sum is odd.
2) One player is called **Low** and the other is called **High**. The lowest product on the dice is 1 and the highest is 36. **Low** wins when the product of the dice is **less than 15**. **High** wins when it is 15 or above.

TRY OTHER INTELLIGENCES • • • SHOW YOUR THOUGHT WORK • • • REPORT YOUR PROGRESS

5. You are working for a pottery teacher at a camp. Your teacher is mixing some new clay that must have the perfect feel and he needs **exactly four** quarts of water, according to his secret recipe. He sends you to the river with the only two containers he has, a couple of pottery **vases** with thin necks. He knows that one holds **exactly 5** quarts. The other holds **exactly 3** quarts. How can you bring back **exactly 4** quarts?

(Remember, _no estimating allowed_, but pouring between vases and the river is!)

There are **two** different pouring sequences that will give you your result. Can you find **both**?

TRY OTHER INTELLIGENCES • • • SHOW YOUR THOUGHT WORK • • • REPORT YOUR PROGRESS

6. A coin collector loves rare coins. A dealer offers to sell her 18 large Roman gold coins for $500 each. She gets a secret tip from someone that one of the coins is counterfeit (not solid gold) and has a core made of lead instead of gold. This makes it just **slightly lighter**, but not enough to be able to feel the difference, or even check it on a cheap balance scale.

Obviously she doesn't want to buy the counterfeit one and lose $500. She has heard of a very sensitive **scale with two pans** that can check for **balance** between two objects to see if they weigh **exactly** the same. She says "great!" and takes her coins there but she is told it will cost $50 every time she uses the scale to **check for balance**. Even if she checks one pile of coins against another pile (for $50) then removes one coin from each pile to check for balance again, that's another $50. She goes "Aha!" and realizes how she can find the counterfeit with the machine for only $150 (three checks for balance). How can she do it?

TRY OTHER INTELLIGENCES • • • SHOW YOUR THOUGHT WORK • • • REPORT YOUR PROGRESS

7. The number 5 __, __ 8 6 is a five digit number that is exactly divisible by 7. So is 6 3, __ __ 2. Can you find digits for the blanks that make this true?

Can you find the same two digits (in either order) that will work to make both numbers divisible by seven?

TRY OTHER INTELLIGENCES • • • SHOW YOUR THOUGHT WORK • • • REPORT YOUR PROGRESS

8. A man and woman are at a small settlement near a hot, sandy desert's edge. They have heard that a friend of theirs is in grave danger. He is being hunted by a bandit, and he'll die if he's not warned before he gets to a faraway village. They don't know where he is exactly, but they must warn him. An idea: They know that sometime between 9 and 40 days from now he will pass a place in the desert that is **nine days walk** away from them! They must leave a big sign for him there.

There is no food or water at the message place so they must carry all supplies (food and water) along with them. Each of them can carry the food and water that will last one person exactly **12 days**. (They can't stretch it out — a whole day's worth of food and water *must* be used each day to keep from dying in the intense heat.) They conclude that only one of them (the messenger) should make the walk so that the other (the assistant) can help. But how? The assistant can **bury** food and water for the other to find on the way back or can **give** them to the messenger on the way out. How can they get the message delivered and return to safety?

Championship Solution: How can the assistant help the messenger to return safely and only enter the desert <u>once</u>?

TRY OTHER INTELLIGENCES • • • SHOW YOUR THOUGHT WORK • • • REPORT YOUR PROGRESS

9. You wish to make a **different** colored **three-color badge**, like the one on Wily's hat, for each of 20 students in your club. *"Different" means that no two badges have exactly the <u>same</u> three colors even if the colors are in a different order on the badge. That is, each badge must differ from all the others by at least <u>one</u> color.* You have **six** pens of different colors. Do you have enough pens to make all the badges? Whether your answer is "yes" or "no," tell how many badges you *could* make and how you figured it out.

TRY OTHER INTELLIGENCES • • • SHOW YOUR THOUGHT WORK • • • REPORT YOUR PROGRESS

10. Try to make the numbers **1 through 25** (or 1 through 50 if you want more challenge) with **five 4s** and any math symbols you want. You get a bonus if you can make it with **four 4s**.

Examples: $9 = 4 + 4 + 4/4$
or $9 = 44 \div 4 - \sqrt{4}$
or $9 = 4/4 + 4\sqrt{4}$
and $2 = 4\sqrt{4} + 4/4$

If you get **stuck** on any numbers you can:

• **discuss** it with others in class;

• give it as a **challenge to adults** or **other kids** around you who are interested in math;

• **post it** in the hall as a challenge in your school — include slips that tell where to submit answers, and even hold a prize-drawing from the names of those who enter correct answers, or

• for the real stumper numbers write a **letter to the editor** of your local paper and ask for solutions to be sent to your school address. (Some answers might even use math ideas you haven't learned yet!)

TRY OTHER INTELLIGENCES • • • SHOW YOUR THOUGHT WORK • • • REPORT YOUR PROGRESS

11. Find **1,000,000 of something** and **prove it**.
(First discuss with someone what you think that means, then do it! Your teacher may give you some ideas about it too.)

TRY OTHER INTELLIGENCES • • • SHOW YOUR THOUGHT WORK • • • REPORT YOUR PROGRESS

12. A woman has a very important and skilled job she wants done. A man tells her he can do it. They figure out that if he works **exactly 15** workdays (from 9 a.m. to 6 p.m.), he will be able to finish the job, BUT she must offer him a good wage. She says, "Indeed I will! I have a bar of gold that is **exactly 15 inches long**! I will pay you **one inch of the length of the bar each day** you work. They agree that she will pay him the inch at 6 p.m. each day. And so it was — he worked the 15 days and she did her payments on time as agreed. She had discovered that she only needed to **cut the bar three times to keep her bargain.** ("Cut the bar" means to lay it flat on something and saw it **straight down** from its top surface to its bottom surface.) How did she do it?

TRY OTHER INTELLIGENCES • • • SHOW YOUR THOUGHT WORK • • • REPORT YOUR PROGRESS

Teacher Notes

Intelligences Emphasized: All

Math Skills, Concepts: Problem-solving strategies, logic, whole-number relationships, estimation, combinations, multiples, area, averages

Materials: Provide counters, cubes, colored pens, ruler, paper, scissors, calculator as options for student use.

Level of Challenge: Moderately challenging (#1, 2, 3, 4, 5); more challenging (#6, 7, 8, 9); very challenging (#10, 11, 12)

Overview: *This Activity is intended to follow, and put into practice, a discussion of the theory of Multiple Intelligences you have already held with your class. You can get the basics of that theory in Chapter 1.* In this Activity, which contains some of my favorite brain-stretchers, you can teach children how to utilize **several intelligences** to solve puzzle-like situations. Remember, a problem is a situational question to which a student has no immediate answer, but which is within reach of his/her skills after some effort. It's not a problem — only an exercise — unless a student STARTS OUT STUCK. Being stuck is not the *end* of the need for effort (as many students have been coddled into presuming) but the *beginning*.

Becoming a good problem-solver means developing **attitudes** like *perseverance* (coming back to the problem again if it resisted before) and *confidence* (a belief that there *is* a solution and that continued effort will find it). Once someone asked Daniel Boone if he was ever lost. After scratching his head a while he said, "Can't say I was ever *lost*...I was *confused* for two months once." This is a great example of a problem-solving attitude.

Problem-solving also means expanding one's range of **strategies** for what to do when stuck. Most math textbooks these days develop at least a limited repertory of such strategies, though they don't embark on the attitude development I mentioned or on broader thinking strategies applicable to a wide range of problems. An example of one of these is, "Get *some* new fact even if it's *not* the answer." Often students get fixated on just arriving at the *final* answer so the tension will be over. This diverts their attention from using the data to find many sub-results that will then cue them *toward* the final answer. The "get-*some*-new-fact" strategy is a good way to get students moving on a brain-stretcher.

What I am emphasizing in *this* Activity, though, is something also not mentioned in textbooks: that **shifting to another intelligence** can often lead to a breakthrough, because we may think better in that intelligence. This can be called another of the broad strategies — applicable to a wide range of problems. How this is to be done is sketched in the student instructions and will be expanded in my discussion of each brain-stretcher below, under **Answers.**

The **Multiple Intelligence Strategies** page and the **Advice for Working the Brain-stretchers** page have many principles involving how to use the intelligences and how to proceed on the brain-stretchers that you may wish to repeatedly emphasize during the brain-stretcher work. As advised there, discourage students from trying to "knock off" all 12 brain-stretchers in a sitting. They can't do them justice. This is a problem banquet, a multi-course feast. A single brain-stretcher can be the source of a day's instructional activity. Even if an answer is hit upon rapidly by a group, this is only a piece of the action. Every student should be prepared to be called on to give the steps of the solution. All students should record notes on their explorations as they go. Reaching a solution (or dead end), they need to report, coherently and in an organized fashion, on the successful (or unsuccessful) process, either in writing or orally. The **light bulb** extension present in several of the brain-stretchers gives you the option of increasing the challenge for the faster groups or more able students.

What is your **role** while the problem-solving process is going on? While you must adjust to being on the sidelines, you will be far from idle. You will be:
- an *observer* of the process of groups and individual students,
- an *evaluator* of the problem-solving development of individuals (see Chapter 8: "Assessment Ideas"),
- a *giver of graduated hints* (if a whole group is stuck for a while),
- an *encourager*,
- a *verifier* of answers, and
- an *open-ended-questioner* who restores movement when a group has "run aground," or when their results are inadequate or incomplete.

Remember, closed-ended questions, like "Which numbers should you multiply here?" almost force a certain preconceived answer from the listener. Open-ended questions don't narrow the focus, they just guide the search. Initial **open-ended questions** to a stuck group, or one with inadequate results or documentation, are:

"Have you tried switching intelligences? Which intelligence is most applicable to what you're doing?"

"Are you making note of what you have gotten so far, even if it's not the answer? Is there any more you could get, even if it's not the answer?"

"Are you stuck in a groove? What can you change in your approach?"

"Could I call on *any* member in your group to explain how you got your answers?"

"Could anyone who has not been in this room easily figure out from your papers how you solved this brain-stretcher?"

"Have you completely answered the question?"

Introduction to Class: NOTE: You need to hold a general discussion of the theory of Multiple Intelligences with your class before you proceed with this Activity (see Chapter 1: "Multiple Intelligences"). Then hand out pg. 14-1 to all. Present the purpose of the Activity (to learn to shift intelligences to create more successful problem solving) then have them read the Multiple Intelligences Strategies sheet and play the "Timed MI Sort Game", both requested on pg. 14-1.

Then hand out page 14-3, with its "Advice for Working the Problems". It would be helpful to have a short class discussion of these before any problems are started. Go back over them if you feel their advice is being neglected.

Finally, choose or vote on a single brain-stretcher from the collection. Discuss it until everyone is sure what questions the brain-stretcher is asking before any attempts at answering them are entertained. Students may be invited to go to work on it in groups or individually. For groups, try to build in the cooperative learning principles discussed in Chapter 6. My experience is that a group of students, where all start out stuck, will solve a problem about three times faster than will even an above-average student doing it alone (cheers for the interpersonal intelligence!). A group report on the process of solution, either orally from one or more members of the group, or in writing from the entire group, will complete the brain-stretcher.

Save the rest of the brain-stretchers for other days of intensive group problem-solving. They can also be used as enrichment for students that are more able, or as "problem of the week" home-solving for the class.

Answers

1. 55 rabbits, 6 hutches. The secret is to make a list of the table of nines and one more than each entry as candidates for the number of rabbits, i.e.,

The 9 multiples: 9, 18, 27, 36, 45, 54, 63, 72, etc.
Add one to each :10, 19, 28, 37, 46, **55**, 64, 73, etc.

When one of the entries is also a multiple of 11 (i.e., 5x11), the solution is found. When the 55 rabbits are placed 11 to a cage, one whole hutch is left over, so there must be six hutches. As for the intelligences, students can model containers and put counters in each to represent rabbits as they explore. Or they can simply draw circles and put rabbit symbols in each until they begin to see a pattern of multiples of 9 plus 1. Reciting the second line rhythmically may give students an idea of extending it while looking for an 11-multiple. The naturalist intelligence can help by sorting and classifying numbers in the lists relating to a solution.

 Nine hutches. **Hint**: The two numbers a student chooses for making up a brain-stretcher will only lead to whole number solution if their *difference* divides evenly into one more than the larger number.

2. Father 4, Daughter 6. This brain-stretcher invites everything from walking it through, to diagramming, to moving objects of two different colors to represent footsteps, using sticks or fingers as walking legs, feeling it as a 3-against-2 rhythm, or dealing abstractly with common multiples of 3 and 2.

Basically, as the father takes two more steps, he will be back on his left. By then the daughter has gone right-left-right, so the feet don't match. So the father takes two more steps — right-left. The daughter takes three — left-right-left — and they both have landed on their left.

 The father will take six steps and the daughter will take 8.

3. There are many possible answers and many methods to get it. One method is to subtract a certain amount, say 10 pounds, from 173, and add 10 to 173 to get 163 and 183, which still average to 173. Do the same again with say 12 pounds, to get 185 and 161 that still average to 173. Now we have a fifth 173. So 173, 163, 183, 161, and 185 average to 173.

jottings

X	1	2	3	4	5	6
1	(1)	2	(3)	4	(5)	6
2	2	4	6	8	10	12
3	(3)	6	(9)	12	(15)	18
4	4	8	12	16	20	24
5	(5)	10	(15)	20	(25)	30
6	6	12	18	24	30	36

Another method involves making up some large numbers and some small numbers then averaging them to see how close to 173 they are. If their average is too high, kids may want to keep adjusting weights down until the average is right. A more scientific way is this: Suppose the *average* (obtained by dividing the total of the tries by 5) is too big by, say, 14 pounds. This means the *total* of the weights is too big by 5x14 = 70 pounds. Remove this much from the various weights in any sized chunks and ...*Voila!*

> **Hint**: "Draw some tall (heavy) and some short (light) stick persons. Draw a line through them that represents the height of 173 pounds. Make numbers for each person's weight. Cut off the tall ones to fill in the short."
>
> **Hint**: "Make five towers, some tall and some short, with stacking cubes. Let each cube stand for 10 pounds. Make them so the average tower length is 17 cubes or 170 pounds. Then adjust the numbers a bit so the average is 173."

4. Not fair. Odd will only win once for every three times Even wins. There will soon be a sense that it isn't fair, but the task is in proving *how* unfair it is. It boils down to looking at all possible outcomes of any play. That is, if a 1 is thrown by the first person, what could be thrown by the other player? A 1, 2, 3, 4, 5, or 6. And what if a 2 is thrown by the first person? And so on. Students may try various worthwhile ways to display all possible outcomes. Probably the most efficient way, which you can propose in a follow-up discussion, is to use a table as shown here. Simply circle the outcomes that can be odd.

Be sure to discuss why there are so many evens: any time we multiply by just one even number, the factor 2 must occur in the answer, making it even. The 2-factor is "contageous."

 1) This is a fair game. A table will demonstrate that there are as many even as odd sums possible. 2) This is not a fair game. 26 of 36 possible sums are below 15 and 13 of 36 are 15 or above. In 100 throws we would expect 26/36 of 100, or 72 points, to go for Low, and 13/36, or 38 points, to go for High.

 5. First way: Pour the 5-vase into the 3-vase (then empty this vase), leaving 2 in the 5-vase. Pour the 2 into the 3-vase. Then fill the 5-vase and pour it into the 3-vase (only one quart will go in). This leaves 4 in the 5-vase. *Voila!*

Second way: Pour the full 3-vase- into the 5-vase. This leaves two empty quart spaces in the 5-vase. So, fill the 3-vase again and pour into the 5-vase again. This time there is 1 quart left in the 3-vase. Empty the 5-vase and pour that 1 quart into it. Now pour a full 3-vase in also, and you have 4 quarts. *Voila!*

> **Hint:** Use stacks of five blue cubes to represent the 5-vase and three yellow cubes for the 3-vase. If the 3-vase is filled from the 5-vase, change it to a blue stack. If the 3-vase pours into the 5-vase, change three of the 5-vase cubes to yellow.
> **Hint:** Ask the student, "What possible things can you do at each stage of your pouring? Try all of the possibilities, and you will see there is only one sensible pour to do to keep going toward a solution."
> **Hint:** Use a decision-tree, codes, or vase pictures to show your possible decisions at each stage.

6. Place six coins on each pan of the balance scale, leaving six aside. Whether or not they balance, you can conclude which group of six coins contains the counterfeit. From that group, put a two coins on each pan. Whether or not they balance, you can conclude which pair of coins contains the counterfeit. Place one of that pair on each side of the scale. The counterfeit is found.

> **Hint:** Be sure to try modeling with real objects, or your hands imitating a balance scale, or a drawing.
> **Hint:** "Try doing something different, like how you group the coins."
> **Hint:** "Do you get information about the coins that are *not* on the scale when you place coins *on* the scale?"

7. There are several answers for the digits.
There are many answers to the first one. We are searching for a factor number, __ __ __ 8, which multiplies by 7 to make the answer 5 __ __ 8 6. (The ending 8 comes from the fact that 8x7=56, which makes the big number end with 6.)

You can reason why the second-to-last digit of the top number is 9 by looking at the multiplication to the right. The first digit of the top number must be 7 or 8 because 7 x 7 = 49, close to 50-something, and 7 x 8 = 56. The end digit is 6 because 8 x 7 = 56 and the leading 5 of this carries. The tens digit 9 is in the top number because 7 x 9 = 63 and the carried 5 makes 68, making 8 in the answer. If students start with the forced numbers of the multiplication above, they can experiment to get the other digit of the top factor and thus the other digits of the answer. It turns

$$
\begin{array}{r}
{\scriptstyle 6\ 5} \\
7\_\,98 \\
\times 7 \\
\hline
5\_\_\,86
\end{array}
$$

out that any numbers 8a98, where a = 0 to 4 or 7u98 where u = 1 to 9 can be multiplied by 7 to get the appropriate three given digits of the answer.

Hints for the second are that the number begins with 63 (9x7=63) and ends with 2 (7x6=42). So we are looking for a number 9__ __ 6 that multiplies by 7 to make 63, __ __ 2. As students experiment with digits, they will arrive at their individual answers, which should all be the answers to multiplying 90a6 by 7, where a = 1 to 9 or 91u6 by 7 where u = 0 to 3.
This is primarily a mathematical-logical exercise helped greatly by interpersonal group discussion and exploration.

Extra sharp students should explore how many numbers they can find to make correct answers as discussed above. If they list all possibilities for answers they will find that 55,986 and 63,952 are divisible by 7 and utilize a 5 and 9 for the missing digits.

8. Championship solution: The assistant (*A*) and messenger (*M*) walk 3 days into the desert. Now they both have 9 days of supplies. *A* gives *M* 3 more days supplies, so *M* has a full 12. *A* also buries 3 days supplies right there, then returns to safety using the remaining 3 days supplies. *M* walks the remaining 6 days of desert, delivers the message, and walks 6 days back to the buried 3 days' supplies. Using those, *M* makes it out of the desert.

There are many solutions students will arrive at that are not the championship solution. Many can be correct though less efficient. You can guide the exploration with early reminders and hints that include these:

Reminders:
"Both people don't need to deliver the message."
"It requires the same number of days of supplies for anyone to get back *out* of the desert as it takes to go *in*.""The assistant can give over supplies to the messenger or bury food for his/her return."
"The assistant can go into the desert any number of times and buy supplies any number of times at the starting place."

Hint: "Act it out. Have one person play the messenger, and another play the assistant. Carry the number of cubes that are your days' supplies of food and water. Make markers on the floor for each days' walk into the desert."
Hint: "Try having the assistant walk into the desert with the

messenger one day, two days, three days, etc. and consider what they can do with their various options each time."

Hint: "There are different difficulties of solutions: If you can have the assistant go into the desert only **three times** that's **good**. Only **two times** is **very good**. Only **one time** is a **championship** solution!"

9. Yes, there are just enough pens. Six pens will make 20 three-color combinations, each differing from the other by at least one color. There are many ways to solve this, but interpersonally, visually and kinesthetically are the best. Using the naturalist intelligence to sort and classify combinations helps too. Actual use of six pens and a group effort is a good way to spot repeating color combinations that any individual might miss. Each time a new combination is suggested, all others must be scanned to see if it is redundant in any order.

One can be more systematic by starting with red, for instance, and getting ten different combinations for the other two colors with the five remaining pens. After these combinations that start with red, start with another color, blue, to get six new combinations only, then a third color for three more combinations, and a fourth color to get one final combination.

If colored cubes or counters or paper squares are used instead, it allows each student to make suggestions by putting forth a set of three. If it holds up to scrutiny then it is included in the group.

> **Hint:** (If students seem to be satisfied with less than 20 or more than twenty as a final list, re-state the brain-stretcher as: "There are exactly 20. Can you find what they are?")

10. Answers will vary. This is an excellent activity for group, whole class, school, and even community involvement. It includes a lesson on how to submit a letter to the editor, a valuable life skill. Post the solutions for various numbers prominently around the room. This could last for a few weeks as students search for tough number solutions among peers, adults, etc. It can even be extended to numbers all the way to 100.

> **Hint:** Develop a repertory of numbers you can construct others from, like $4/4 = 1$, $\sqrt{4} = 2$, $4 - 4 = 0$, then try to put them together to get your number, i.e., $(4-4) \times 4 + 4 = 4$. You can decide to use 4^2 or 4^3 (since, technically, squaring and cubing are operations) if you wish. You can also introduce $4!$, meaning $4 \times 3 \times 2 \times 1 = 24$ (the "!" key can be found on scientific calculators).

11. Answers will vary. This is called an open-ended problem. The teacher doesn't know the answer or the method, and the student, initially, doesn't either. Work in pairs or alone can be good here. Some "convergent" students (beavers) will beg for more guidelines while "divergent" thinkers (foxes) will feel liberated. Ideally, "prove it" means to show actual calculations that will persuade me that you really have about a million. Depending on how much autonomy and divergent thinking you have fostered in the students, you should try not to pin it down for them — leave it open to their imagination to figure out what the assignment means and how to do it.

After a first round of this activity as class or homework, discuss everyone's results. Then come up with more restrictive guidelines and assign it again. For instance, request 1,000,000 *concrete discrete objects* that you could *touch* and *see* individually (like blades of grass or stitches in a sweater, rather than the 1,000,000 millionths of a foot in a ruler or 1,000,000 atoms in a pencil point).

Also, in the second round, refuse to accept vague estimates like, "There are at least 1,000,000 rocks in the pavement in front of the school." The answer must be specific, like "In a 14-foot by 50-foot section of the pavement directly in front of the school entrance there are 1,000,000 rocks size 1/4" or larger. This is because a 4" by 4" square has 120 rocks, and I multiply that by 9 to get the rocks in one square foot. Then I divide 1,000,000 by that figure, which tells me I will need 694 square feet or a 14 by 50-foot area to get a million rocks."

12. She cuts the bar with three cuts into four pieces that are 1", 2", 4", and 8" long, then pays and "gets change" each day. On the first day she pays 1". On the second she gives 2" but asks for the 1" back. On the third day she pays the 1" again, and so forth.

This is a good one for students to work on alone, though groups of three or four will work well also. Alert them in the beginning: "Be very discrete about your proposed answer, especially if you have experienced a major Aha! (If you don't have a strong, certain feeling that the answer is right, it probably isn't.) Do not blab it out, but instead come and whisper it into my ear. If it's correct, guard the secret with your life."

"It's a giant no-no to short-circuit someone else's Aha! pleasure by telling them the answer! Though others may seem to want relief from the unknown, the pleasure of being told the answer is very short-lived and is only about 1/50 the pleasure of getting it themselves — even if they have to get a few hints from me. So don't rob them of this treat by telling them. Later on you will get your recognition for finding the right answer."

And don't feel it has to be answered *that day*. One major principle of problem solving is that we need to *incubate* the solution by leaving it alone for a while (even for a night's sleep). When we return we find we have a fresh perspective. (I've had students report that they had breakthroughs on problems while doing other non-math tasks.)

Hints (In increasing order of specificity):

Hint: "A 'cut' of the bar is simply straight through, not diagonal or wavy."

Hint: "How many pieces will anyone get when they make *three* cuts in a bar?" [four]

Hint: Take a piece of paper, or 15 cubes stuck together, and try possibilities for cuts. Act it out with someone, day by day.

Hint: Payments are not in groups, they occur each day at 6 p.m. and no credit is granted by either person beyond the one day's work.

Hint: "Think one day at a time. What will happen on the first day? [She will have to cut one inch for that payment.] The second day? This is the crucial decision day. Why? Try everything you can except what you obviously want to do." [She need not cut another inch for payment. She can cut two inches and trade.]

Hint: Don't get hung up on the idea that 15 payments mean 15 pieces. How can more payments be done with less pieces? Remember that the man still has his pieces of gold after she pays him.

Hint: Think about the word TRADING.

Extension to brain-stretcher 12: Base 2. It is no coincidence that the numbers 1, 2, 4, 8 have a pattern. These are the powers of two and are used in computers to make all numbers as sums of these. So 1, 2, 4, and 8 can make all numbers (payments) up to 15. Similarly, if there were a 31-inch bar of gold, cuts that make 1, 2, 4, 8, and 16 would make all numbers up to 31. Explore with the binary numbers. What is the next level of this puzzle? [63" bar; 5 payments]

jottings

Extensions:

Brain-Stretcher Search. Students research to find other "favorite brain-stretchers". There are many sources on the internet. See www.markwahl.com for some good links.

Champ Problem-Solvers. Students nominate people from all walks of life — living, dead, or fictitious — as good examples of problem-solvers (e.g., Geronimo, Nancy Drew, Bill Gates, the car mechanic). They say what qualities their nominee has that make him or her a good problem-solver.

Model Solutions and Strategies. Students critique the solution write-ups of any brain-stretcher in their groups and eventually (when trust is built) as a whole class. These are polished until they become models for making quality reports in the future. These model solutions can also be posted in the hall. Each time a new strategy proves useful for students it is discussed, summarized in a few words, and posted in the room for future reference in solving problems.

RESOURCE BIBLIOGRAPHY

Listed here are catalogs, books, videos, and manipulatives that may be useful as you expand your techniques. You may have been referred here by a note in one of the chapters. Each resource has a short commentary and code or phone number(s) for obtaining it. This is by no means an exhaustive list, but rather one that gives you a starting point for expanding in some of the areas and techniques that appeal to you as you read *Math for Humans*. Updated January 2008.

USEFUL CATALOGS

CP   Creative Publications. Large assortment of math learning aids, manipulatives, and books as well as integrated curricular resources. (888) 205-0444. Click the icon at creativepublications.com.

ET   ETA/Cuisenaire. Primarily a math manipulative company with teacher books, software, and science materials mixed in. (800) 445-5985. etacuisenaire.com.

NCTM National Council of teachers of Mathematics. Numerous readable books and some teacher journals addressing many aspects of teaching most math concepts at all levels. (800) 235-7566. nctm.org.

PP   Prufrock Press. Gifted education, thinking skills, puzzles, simulations. Many subjects. (800) 998-2208. www.prufrock.com.

CWP   Corwin Press. Large collection all subjects. (800) 233-9936. corwinpress.com.

MCL   The Math Learning Center. Math resources and manipulative materials. Colorful catalog or online ordering. 800-575-8130. mathlearningcenter.org.

MANIPULATIVES (For more discussion of each, refer back to "Seasoning with the Bodily-Kinesthetic Intelligence" in Chapter 3: "Seasoning Math with MI." Following each description here you will find the code(s) of the mail order house(s) giving the best selection.

Multilink® Cubes. They connect in all directions and have ten colors. Triangular pieces join the cubes to create many polygons. There are transparent versions for the overhead. Various activity books in patterns, puzzles, fractions, geometry, and pre-algebra are available. ET

Base Ten Blocks. Unit cubes, tens rods, hundreds flats, and thousands large cubes model place value of numbers and the operations of addition and subtraction with regrouping. They model simple multiplication and division ideas also. A wide variety of books of activities and explorations can be purchased to go with them depending on your curricular needs. ET

Decimal Factory™ Blocks. These extend the Base Ten Blocks for use with instruction of decimals. They come with an activity binder that encourages hands-on explorations. CP

Cuisenaire® Rods. Ten colors and ten lengths model the numbers 1 through 10. Two 10s and a 3 laid as a "train" model 23. Number patterns, addition facts, multiplication facts, factors, as well as addition, subtraction, and multiplication algorithms are especially easy to model with these. Various activity books are available also. Be sure to purchase meter sticks (see the discussion in Chapter 3 under "Seasoning with the Bodily-Kinesthetic Intelligence") to expand the symbol power of the rods. ET

Fraction Bars®, Fraction Pieces, and Fraction Tiles. These create different helpful models for many aspects of fractions. Often multiple models are better than a single one. All three models come with various manuals and activity books. CP, ET

Pattern Blocks. These multi-colored geometrically-shaped blocks come in a variety of set sizes, in both plastic or wood. (I recommend the wood for a better tactile experience.) Each company I've listed has a variety of books of pattern block activities and teaching theory that you can choose from according to your needs. ET.

Wooden or low-noise foam cubes. These make large or small dice with various numbers or equations on the faces. MLC

Transparent four-colored plastic tokens. These model place value chip-trading and cover numbers on grids. MLC

BOOKS AND VIDEOS (Obtain them through the phone number, website, or company code at the end of each description or, if none is given, search web book sellers.)

S. Beall, *Functional Melodies*. Activities for intermediate and higher that link music and mathematics, including "sound shapes," time measures, and hearing number relationships. Key Curriculum Press, 800-995-MATH, www.keypress.com.

J.G. Brooks and M.G. Brooks, *In Search of Understanding: The Case For Constructivist Classrooms*. Association for Supervision and Curriculum Development, (800) 933-2723. www.ascd.org. Get their catalog of many teacher resources.

M. Burns, *About Teaching Mathematics: A K-8 Resource*. Marilyn Burns tends to emphasize logical-mathematical development through problem-solving. This book contains many imaginative activities, teaching theory, bibliography, and blackline masters. ET

M. Burns, *A Collection of Math Lessons. Grades1-3 or Grades 3-6 or Grades 6-8*. With a strong emphasis on problem-solving thinking, they have many imaginative lessons that can be integrated with the year's curriculum. ET

M. Burns, *50 Problem-Solving Lessons*. A compilation of classroom-tested suggestions in various math strands from teachers across the country. ET

M. Burns, *Writing in Math Class*. It justifies the use of writing in math class and delineates five different types of writing assignments, showing samples of student work. MarkWahl.com, ET

L. Campbell and B. Campbell, *Multiple Intelligences and Student Achievement: Six Success Stories from Elementary, Middle School and High School*. ascd.org

D. Campbell and C. Brewer, *Rhythms of Learning: Creative Tools for Developing Lifelong Skills*. A practical presentation on the many aspects of rhythm in the learning process: from emotional and body rhythms to schedule rhythms to use of music in the classroom. Zephyr Press

Davis, Ron, *The Gift of Dyslexia*. Gives deep insight into how dyslexia works to subvert math and gives a proven remedy for the condition. For more see Ch.4 of this book; also Freed entry below. See www.dyslexia.com for moch more information.

Freed, Jeffery, *Right-Brained Children in a Left- Brained World*. Deals with the ADD and ADHD diagnosis explosion, and the traits and internal life of ADD youths. He claims they are quite right brained with low impulse control. He recommends alternative teaching approaches. (for more see Ch4 of this book and see Davis above for possibility ADD diagnosis masks dyslexia.)

Howard Gardner, *Frames of Mind: The Theory of Multiple Intelligences*. The original ground-breaking, readable, treatise on MI theory. Good for going into more depth on why these were chosen and details of how each intelligence manifests.

Howard Gardner, *Multiple Intelligences: The Theory in Practice*. A good theoretical basis for using the Multiple Intelligences in the classroom. Also, a good discussion of more enlightened modes of assessment.

W. Glasser, M.D., *The Quality School* and *The Quality Teacher*. Short, readable books that present a model of classroom management that replaces "boss-management" with "lead-management." Glasser has adapted the highly successful Deming management model used in business to the educational setting. It always attempts to provide work that satisfies the five basic needs of every human being: love, power, freedom, fun, and survival. The emphasis is on self-evaluation and striving for "quality" in all work.

N. Gordon, *Magical Classroom*. A book and audiotapes related to classroom visualization as an aid to teaching.

A. Gregorc, *An Adult's Guide to Style*. Gives the four learning styles used in **Math for Humans** in a more schematic way. Gabriel Systems, Inc. www.gregorc.com

C. Hannaford, *Smart Moves: Why Learning is Not All In Your Head* and *The Dominance Factor*. The first book is the best place to start to learn how movement is intimately related to learning. Both are highly informative about learning strengths and difficulties. Explains how right/left dominance of the brain, hands, feet, eyes and ears influence learning, and how to enlist kinesthetic intelligence for remediation.

C. Hannaford, C. Shaner, S. Zachary, *Education in Motion* (Video). Researched and tested exercises that increase ability to learn in any subject. Adaptable to math class.

O. Ho, *Amazing Math Magic: Easy Magic Tricks with Numbers*. For youths.

B. Irvin, *Geometry and Fractions with Pattern Blocks*. This book has hands-on explorations of geometric principles and fraction relationships for Grades 3-6. ET

E. Jensen, *Brain-Based Learning*. An almost encyclopedic summary of recent findings and theories about how we can learn better. Jensen Learning Corporation. www.thebrainstore.com. (Get a catalog of many other accelerated-learning and brain-compatible classroom strategies available.)

D. Johnson, R. Johnson, E. Holubec, *The New Circles of Learning*. A short, readable rationale and methodology for using the cooperative learning model in the classroom. Gives many nuances and variations learned from extensive research and experience. Interaction Book Company, www.co-operation.org (Many cooperative learning papers and booklets available).

D. Johnson, R. Johnson, *Learning Mathematics and Cooperative Learning*. Math activities from veteran teachers usable with primary, intermediate, junior and senior high school cooperative groups. Interaction Book Company, www.co-operation.org (Many cooperative learning papers and booklets available).

A. Kleiman, D. Washington, *It's Alive* and *It's Alive... and Kicking*, Gr. 3-9. Challenging, stimulating math problems that gross out and entertain youths. PP

R. Knauff, *Short Stories from the History of Mathematics*. Gr. 7 and up. Many anecdotes about significant events and the important historical figures of math. Brief bios of famous mathematicians. Supplemental activities included. Carolina Mathematics, 800-334-5551. www.carolina.com.

D. Lazear, *Eight Ways of Teaching* and *Pathways of Learning*. Gives ways to help students understand their MI learning potentials in four levels of depth: tacit, aware, strategic, and reflective. Helps students develop study strategies and suggests multiple ways to present information to students.

N. Margulies, *Mapping Inner Space*. How to create Mind Maps on any subject. CWP

N. Margulies, *Map It!* A how-to, and rationalization for, mind-mapping in comic-book form for students of middle grades and above. Zephyr Press

D. Markova, *The Open Mind*. Dr. Markova gives a deep and practical theory of how the auditory, visual, and kinesthetic channels create six different styles of operating, depending on which of these is the prime mode of the conscious, subconscious, and unconscious mind. A real help in figuring out how to approach unusual learners.

Mozart Makes You Smarter. A CBS Masterworks CD collection of Mozart's music that especially enhances relaxation and more acute thinking. ZP

R. Myers, *Cognitive Connections: Multiple Ways of Thinking With Math*. Fairly verbal multiple intelligence activities with math skills sprinkled in — some open-ended conceptual, and some with fairly quick calculations. Middle grades. Zephyr Press

National Council of Teachers of Mathematics, *Assessment Standards for School Mathematics*. Gives examples, approaches, overview, theory, and standards for more useful and realistic math assessment in schools. NCTM

National Council of Teachers of Mathematics, *Curriculum and Evaluation Standards for School Mathematics*, and the new 2000 revision of them, *Principles and Standards for School Mathematics*. (The latter can be obtained from the website.) See Chapter 2: "The NCTM Standards" for details of why you should have this on your shelf. NCTM

National Council of Teachers of Mathematics, *Historical Topics for the Mathematics Classroom*. Historical overviews of numbers, numerals, computation, geometry, patterns, and topics from higher math. NCTM

R. Ornstein, *The Right Mind: Making Sense of the Hemispheres*. In it he states that, "the division of the mind is profound," with deep roots in evolution, embryonic development, and society. Ornstein updates his theories of left and right brain in this readable book.

J. Schielack and D. Chancellor, *Uncovering Mathematics with Manipulatives and Calculators, Level 1 and Level 2*. Activities in four strands that emphasize problem-solving, reasoning, communication, and connections. ET

H. Schoen, *Estimation and Mental Computation*. A classic NCTM yearbook presenting several aspects of these valuable life skills and how to involve a class with them. NCTM

D. Seymour, *Line Designs* and *Creative Constructions*. Clear instruction on how to create geometrical designs with only a ruler in the first book and with a compass and ruler in the second book. Good for motivating the "dolphin" learner and those with spatial intelligence. Numerous examples.

L. K. Silverman, *Upside-Down Brilliance: The Visual Spatial Learner*. One-third of the population thinks in images. This book is for those who "think differently" to understand their gifts and mental dynamics.

Jean Stenmark, ed., *Mathematics Assessment: Myths, Models, Good Questions and Practical Suggestions*
The latest methods of math assessment as developed and implemented by various teachers. NCTM

B. M. Vitale, *Unicorns are Real: A Right-Brained Approach to Learning*. Provides sixty-five practical lessons to develop right-brained tendencies in children. A classic best-seller.

Mark Wahl, *A Mathematical Mystery Tour: Higher Thinking Math Tasks*. For fifth grade and up. An interwoven set of explorations that teach number pattern, decimal skills, measurement, and geometry. The guiding concepts are the Golden Ratio and Fibonacci Numbers whose mysterious dance takes the learner from pine cones to the moon and beyond. Mark Wahl Learning Services, 360-221-8842 or www.markwahl.com, ET, PP.

Mark Wahl, *Math Nuggets: 80 Thoughtful One-Page Activities for Pleasure, Insight and Challenge*. These one-shot hands-on activities can be sprinkled into any curriculum to deepen thinking. They are arranged roughly from easier (generally gr. 2-4) to the more challenging (generally gr. 4-8). (All can be made easier or more challenging.) An Answer Key with helpful comments on each activity is at the back of the book. Mark Wahl Learning Services, 360-221-8842 or www.markwahl.com

J. Westley, *Puddle Questions* (Separate booklets for each of the grades 1 through 8). These booklets contain classroom-tested, provocative, open-ended problem-solving questions at grade-level competency and the means to rate them with a scoring rubric. CP

WEBSITES OF INTEREST RELATED TO THIS BOOK

Howard Gardner's website is at **Project Zero** of Harvard, a group devoted to reform in education. Click here to find out about the Multiple Intelligences theory, Gardner (Click "Principal Investigators"), and educational reforms of all kinds (click "Research Projects"). URL: http://pzweb.harvard.edu Also learn about activities in Project SUMIT (Schools Using Multiple Intelligence Theory): http://pzweb.harvard.edu/sumit/

New Horizons for Learning -- The Building
This site (a virtual building) is a must for an incredible variety of information for the growing educator, including an online newsletter, links, articles, reviews, interviews, and breaking information. URL: http://www.newhorizons.org

Check the Mark Wahl Learning Services website for information on math learning and more useful books: **MarkWahl.com**